MT-122, AMG-2

CALCULUS - II

[2 Credits]
Mathematics : Paper-II
For
First Year B.Sc. and B.A. : Semester-II
New Syllabus as per CBCS Pattern
from June 2019

M. D. Bhagat
Ex-Head Dept. of Mathematics
Tuljaram Chaturchand College
Baramati (Dist. Pune)

R. S. Bhamare
Ex-Head of P.G. Dept. of Mathematics
New Arts, Science & Commerce College
Ahmednagar

N. M. Phatangare
Assistant Prof. Dept. of Mathematics
Fergusson College, (Autonomous),
Pune 411 004

Dr. S. G. Purane
Associate Prof. & HOD of Mathematics
PES's Jamkhed Mahavidyalaya
Jamkhed, Ahmednagar 413 201

Dr. A. S. Khairnar
Assistant Professor & HOD of Mathematics
MES's Abasaheb Garware College,
Pune 411 004

S. D. Manjarekar
Assistant Professor
Department of Mathematics
MGV's Loknete Vyankatrao Hiray
Arts, Science & Commerce College, Nashik 422 003

N5037

F.Y.B.Sc. Calculus - II Maths P-II Sem. II　　　　**ISBN 978-93-89533-70-5**

First Edition : October 2019
© : Authors

Published By:
NIRALI PRAKASHAN
Abhyudaya Pragati, 1312, Shivaji Nagar
Off J.M. Road, PUNE – 411005
Tel - (020) 25512336/37/39, Fax - (020) 25511379
Email : niralipune@pragationline.com

➤ DISTRIBUTION CENTRES

PUNE

Nirali Prakashan
(For orders within Pune)
: 119, Budhwar Peth, Jogeshwari Mandir Lane, Pune 411002, Maharashtra, Tel : (020) 2445 2044, Mobile : 9657703145
Email : niralilocal@pragationline.com

Nirali Prakashan
(For orders outside Pune)
: S. No. 28/27, Dhayari, Near Asian College Pune 411041
Tel : (020) 24690204; Mobile : 9657703143
Email : bookorder@pragationline.com

MUMBAI

Nirali Prakashan
: 385, S.V.P. Road, Rasdhara Co-op. Hsg. Society Ltd., Girgaum, Mumbai 400004, Maharashtra;
Mobile : 9320129587 Tel : (022) 2385 6339 / 2386 9976,
Fax : (022) 2386 9976
Email : niralimumbai@pragationline.com

➤ DISTRIBUTION BRANCHES

JALGAON

Nirali Prakashan
: 34, V. V. Golani Market, Navi Peth, Jalgaon 425001, Maharashtra, Tel : (0257) 222 0395, Mob : 94234 91860;
Email : niralijalgaon@pragationline.com

KOLHAPUR

Nirali Prakashan
: New Mahadvar Road, Kedar Plaza, 1st Floor Opp. IDBI Bank, Kolhapur 416 012, Maharashtra. Mob : 9850046155;
Email : niralikolhapur@pragationline.com

NAGPUR

Nirali Prakashan
: Above Maratha Mandir, Shop No. 3, First Floor, Rani Jhanshi Square, Sitabuldi, Nagpur 440012, Maharashtra Tel : (0712) 254 7129;
Email : niralinagpur@pragationline.com

DELHI

Nirali Prakashan
: 4593/15, Basement, Agarwal Lane, Ansari Road, Daryaganj Near Times of India Building, New Delhi 110002
Mob : 08505972553, Email : niralidelhi@pragationline.com

BENGALURU

Nirali Prakashan
: Maitri Ground Floor, Jaya Apartments, No. 99, 6th Cross, 6th Main, Malleswaram, Bengaluru 560003, Karnataka;
Mob : 9449043034
Email: niralibangalore@pragationline.com

Other Branches : Hyderabad, Chennai

niralipune@pragationline.com　|　www.pragationline.com

Also find us on 🅵 www.facebook.com/niralibooks

Preface ...

We have great pleasure in presenting this text book on **CALCULUS - II** to the students of F.Y.B.Sc. and B.A. Semester - II, Mathematics Paper - II. This book is written strictly according to the new revised syllabus of Savitribai Phule Pune University to be implemented from June 2019.

We have taken utmost care to present the matter systematically and with proper flow of mathematical concepts. We begin the Chapter by Introduction and at the end the Summary of the Chapter is provided. We have added one significant feature: **"Think Over It"** in this new **edition**. Here, we have posed questions of simple, difficult and intuitive type in nature. It is expected that the students should think over it and try to find the answers. This will assess the understanding of the knowledge of the Chapter.

The book contains good number of solved problems and the number of graded problems in the exercises.

We are thankful to **Shri Dineshbhai Furia, Shri Jignesh Furia,** Mrs. Anagha Medhekar (Proof Reading and Co-ordination), Mr. Ilyas Shaikh, Mrs. Anjali Muley (Fig. Drawing) and the staff of Nirali Prakashan for the great efforts that they have taken to publish the book in time.

We welcome the valuable suggestions from our colleagues' and readers for the improvement of the book.

PUNE   **AUTHORS**

OCTOBER 2019

Syllabus ...

1. Differentiation (10 Lectures)

1.1 The Derivatives :

Definition of the derivative of a function at a point, every differentiable function is continuous, Rules of differentiation, Caratheodary's theorem (without proof). The chain rule, Derivative of inverse function (without proof, only examples)

1.2 The Mean Value Theorems :

Interior extremum theorem, Mean Value theorems and their Consequences, Intervals of increasing and decreasing of a function, First derivative test for extrema.

2. L'Hospital Rule and Successive Differentiation (10 Lectures)

2.1 L'Hospital Rule : Indeterminate forms, L'Hospital Rules (without proof)

2.2 Taylor's theorem : Taylor's theorem and Maclaurin's theorem with Lagrange's form of remainder (without proof)

2.3 Successive Differentiation : The nth derivative and Leibnitz theorem for successive differentiation.

3. Ordinary Differential Equations (8 Lectures)

3.1 Linear first order equations

3.2 Separable equations

3.3 Existence and Uniqueness of solutions of non-linear equations

4. Exact Differential Equations (8 Lectures)

4.1 Transformation of non-linear equations to separable equations

4.2 Exact differential equations

4.3 Integrating factors

✍ ✍ ✍

Contents ...

1. **Differentiation**   1.1 – 1.36

2. **L'Hospital Rule and Successive Differentiation**   2.1 – 2.46

3. **Ordinary Differential Equations**   3.1 – 3.76

4. **Exact Differential Equations**   4.1 – 4.40

    Appendix   A.1 – A.6

    Model Question Papers   P.1 – P.4

✍ ✍ ✍

Chapter **1** ...

Differentiation

Joseph-Louis Larange

Joseph-Louis Lagrange (25 January 1736 – 10 April 1813) was an Italian mathematician and astronomer born in Turin, Piedmont, who lived part of his life in Prussia and part in France. He made significant contributions to all fields of analysis, number theory, and classical and celestial mechanics.

Lagrange was one of the creators of the calculus of variations, deriving the Euler–Lagrange equations for extrema of functionals. He also extended the method to take into account possible constraints, arriving at the method of Lagrange multipliers. Lagrange invented the method of solving differential equations known as variation of parameters, applied differential calculus to the theory of probabilities and attained notable work on the solution of equations. He proved that every natural number is a sum of four squares. In calculus, Lagrange developed a novel approach to interpolation and Taylor series. He studied the three-body problem for the Earth, Sun and Moon (1764) and the movement of Jupiter's satellites (1766), and in 1772 found the special-case solutions to this problem that yield what are now known as Lagrangian points.

1.1 The Derivative

1.1.1 Derivability at a Point

Let f be a real valued function defined on an interval [a, b]. Then f is said to be derivable at an interior point c, (where a < c < b) if,

$$\lim_{x \to c} \frac{f(x) - f(c)}{x - c} \text{ exists.}$$

If limit exists, the value of the limit is called the derivative or the differential coefficient of the function at x = c and is denoted by f$'$(c).

Hence,
$$f'(c) = \lim_{x \to c} \frac{f(x) - f(c)}{x - c}$$

Left-hand and Right-hand Derivative

We know that limit exist when the left hand and the right hand limits exists and are equal.

Left-hand Derivative at $x = c$:

The left hand derivative of f at $x = c$ is defined as

$$f'_-(c) = \lim_{x \to c^-} \frac{f(x) - f(c)}{x - c} \text{ if this limit exists.}$$

Right-hand derivative at $x = c$:

The right-hand derivative of f at $x = c$ is defined as

$$f'_+(c) = \lim_{x \to c^+} \frac{f(x) - f(c)}{x - c} \text{ if this limit exists.}$$

Thus the derivative f'(c) exist when

$$f'_-(c) = f'_+(c)$$

Example : Consider a function $f_1(x) = x$ and $c \in \mathbb{R}$.

Then $\lim\limits_{x \to c} \dfrac{f_1(x) - f_1(c)}{x - c} = \lim\limits_{x \to c} \dfrac{x - c}{x - c} = \lim\limits_{x \to c} 1 = 1.$

Hence, the derivative of f_1 at c is $f'_1(c) = 1$.

Thus, at every point of $\mathbb{R}$, the derivative of f_1 is 1.

Now, if $f_n(x) = x^n$, $n \in \mathbb{N}$, then $\lim\limits_{x \to c} \dfrac{f_n(x) - f_n(c)}{x - c}$

$$= \lim_{x \to c} \frac{x^n - c^n}{x - c} = \lim_{x \to c} \frac{(x - c)(x^{n-1} + x^{n-2}c + \ldots + c^{n-1})}{x - c}$$

$$= \lim_{x \to c} (x^{n-1} + x^{n-2}c + x^{n-3}c^2 + \ldots + xc^{n-2} + c^{n-1})$$

$$= nc^{n-1}$$

Thus, at every $x \in \mathbb{R}$, $f'_n(x) = nx^{n-1}$.

Illustrative Examples

Example 1.1 : *Show that* $f(x) = x^2 \sin\dfrac{1}{x}$, $x \neq 0$

$$= 0 \qquad , \quad x \neq 0$$

is once differentiable but not twice.

Solution : We have, $f(x) = x^2 \sin\dfrac{1}{x}$, $x \neq 0$

$$\therefore \qquad f'(x) \;=\; 2x \sin\frac{1}{x} \;-\; \cos\frac{1}{x} \quad \text{at } x \neq 0$$

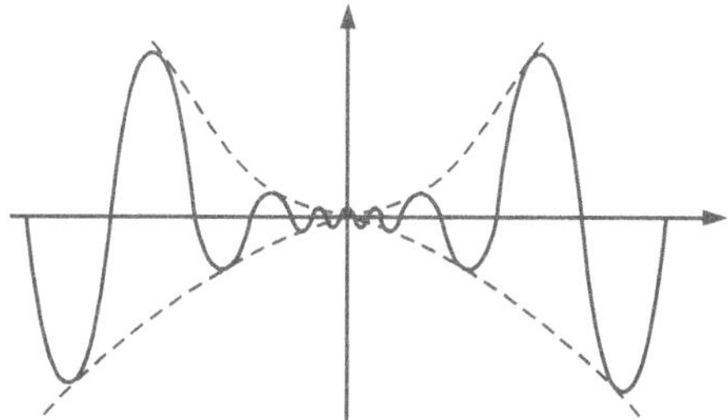

Fig. 1.1

Now the function $f'(x) = 2x \sin\dfrac{1}{x} - \cos\dfrac{1}{x}$ is not continuous at $x = 0$.

$$\therefore \quad \lim_{x \to 0+} f'(x) \;=\; \lim_{x \to 0+}\left\{ 2x \sin\frac{1}{x} - \cos\frac{1}{x}\right\} \text{ does not exist.}$$

Similarly, $\displaystyle\lim_{x \to 0-} f'(x)$ does not exist.

Therefore, the function $f'(x)$ is not continuous at $x = 0$, hence not differentiable at $x = 0$. i.e. $x^2 \sin\dfrac{1}{x}$ is not twice differentiable.

1.1.2 Derivability in an Interval

A function f defined on $[a, b]$ is said to be derivable at $x = a$ if $\displaystyle\lim_{x \to a^+} \frac{f(x) - f(a)}{x - a}$ exists.

$$\text{i.e.} \qquad f'(a) \;=\; \lim_{x \to a^+} \frac{f(x) - f(a)}{x - a}$$

A function f defined on $[a, b]$ is said to be derivable at $x = b$ if $\displaystyle\lim_{x \to b^-} \frac{f(x) - f(b)}{x - b}$ exists.

$$\text{i.e.} \qquad f'(b) \;=\; \lim_{x \to b^-} \frac{f(x) - f(b)}{x - b}$$

The function f is said to be derivable in the **open interval** (a, b) if it is derivable at all points of an interval except the end points.

The function f is said to be derivable in the **closed interval** $[a, b]$ if it is derivable in the open interval (a, b) and also at the end points a and b.

Sometimes, we denote the derivative of f at c by $\dfrac{d\,f(c)}{dx}$, $\dfrac{d\,f(x)}{dx}\bigg|_{x=c}$.

There are other notations that are used for f', such as Δf or $\dfrac{d}{dx}(f)$ or $\dfrac{df}{dx}$ or $\dfrac{dy}{dx}$.

Geometrical and Physical Significance of Derivative :

Consider graph of function f on [a, b] as shown in Fig. 1.2.

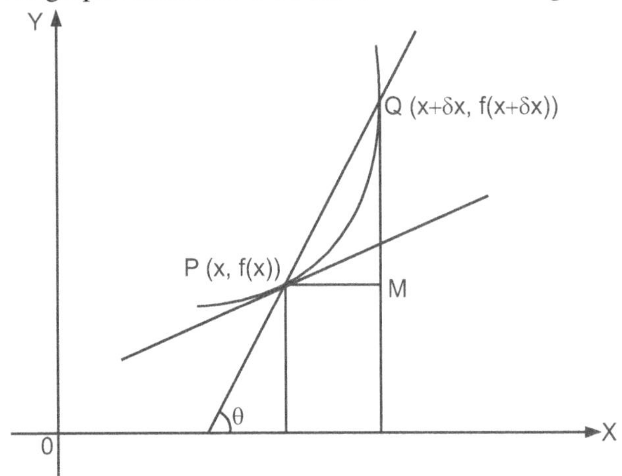

Fig. 1.2

Let P(x, f(x)) and Q(x + δx, f(x + δx)) are two points on the graph of function f. If θ is the angle made by PQ with X-axis then slope of

$$PQ = \tan\theta = \frac{QM}{PM} = \frac{f(x + \delta x) - f(x)}{\delta x}.$$

The limiting value of $\tan\theta$ as Q tends to P (i.e. $\delta x \to 0$) gives us the slope of the tangent at P. Thus, the slope of the tangent at

$$(x,\ f(x)) = \lim_{\delta x \to 0} \frac{f(x + \delta x) - f(x)}{\delta x}$$ i.e. *if tangent line at a point exists then the slope of tangent line is the derivative of f at that point.*

Differentiability Implies Continuity

$\boxed{\textbf{Theorem 1}}$ If the function $f : A \to \mathbb{R}$, is differentiable of c, then it is continuous at c.

Proof : Suppose f is derivable at c, then

$$\lim_{x \to c} \frac{f(x) - f(c)}{x - c}$$ exists, and it is equal to $f'(c)$.

We write, $\quad f(x) - f(c) = \dfrac{f(x) - f(c)}{x - c}(x - c)$ $\qquad\qquad \forall\, x \neq c$

$$\therefore \quad \lim_{x \to c}(f(x) - f(c)) = \lim_{x \to c}\dfrac{f(x) - f(c)}{x - c}(x - c)$$

$$= f'(c) \cdot 0 \qquad (\because \text{ both limits on RHS exist})$$

$$= 0$$

i.e. $\qquad \lim_{x \to c} f(x) = f(c)$

i.e. $f(x)$ is continuous at $x = c$. ∎

Corollary : If f is derivable at all points of an interval, then it is continuous in the interval.

Note : From above theorem it is clear that derivability at a point $\Rightarrow$ continuity at that point but the converse is not true.

For example, (i) If $f(x) = |x|,\ \forall\, x \in R$.

This function is continuous at $x = 0$ but not differentiable at $x = 0$.

Since $\qquad \lim_{x \to 0^+}\dfrac{f(x) - f(0)}{x - 0} = \lim_{x \to 0^+}\dfrac{x}{x} = 1,$

and $\qquad \lim_{x \to 0^-}\dfrac{f(x) - f(0)}{x - 0} = \lim_{x \to 0^-}\dfrac{-x}{x} = -1$

$\therefore \qquad \lim_{x \to 0^+}\dfrac{f(x) - f(0)}{x - 0} \neq \lim_{x \to 0^-}\dfrac{f(x) - f(0)}{x - 0}$

Hence, $\lim_{x \to 0}\dfrac{f(x) - f(0)}{x - 0}$ does not exist.

$\therefore \quad f'(0)$ does not exist.

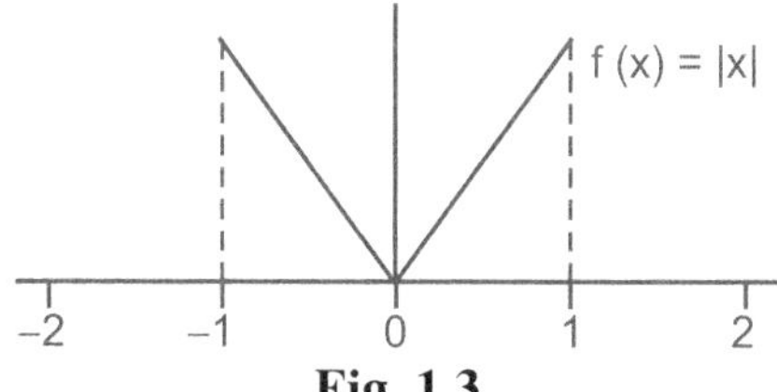

Fig. 1.3

(ii) If $f(x) = |x - 1|,\ x \in [0, 2]$; $f(x)$ is continuous at $x = 1$, but not differentiable at $x = 1$.

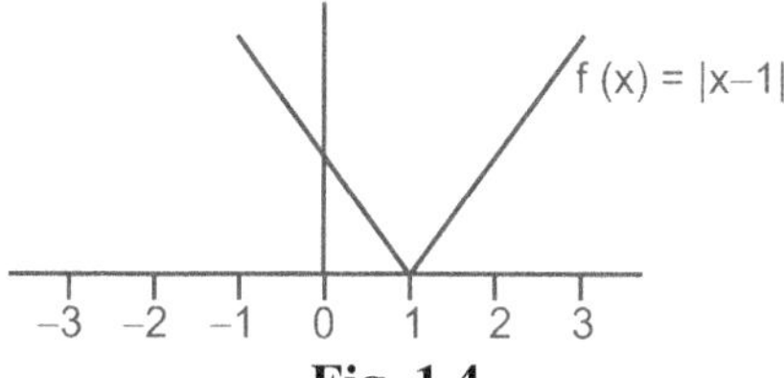

Fig. 1.4

Illustrative Examples

Example 1.2 : *Let* $f(x) = \sin\dfrac{1}{x}$, $x \neq 0$

$\qquad\qquad\qquad\quad = 0$, $x = 0$

is not continuous and not differentiable at x = 0.

Solution : We know that $\displaystyle\lim_{x \to 0} \sin\dfrac{1}{x}$ does not exists.

Thus the function is discontinuous at x = 0, hence not differentiable at x = 0.

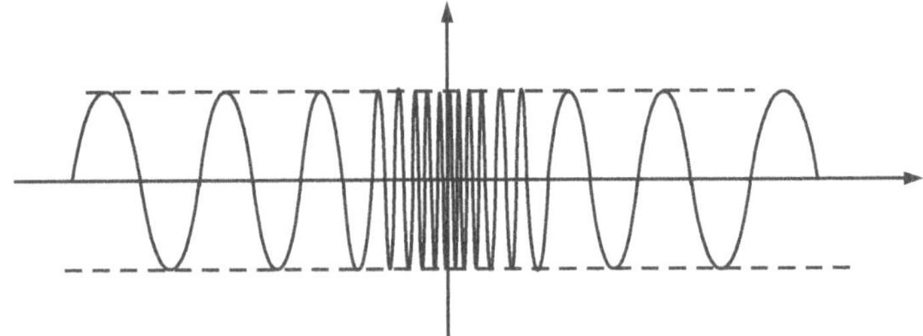

Fig. 1.5

Example 1.3 : *Show that* $f(x) = x\sin\dfrac{1}{x}$, *if* $x \neq 0$

$\qquad\qquad\qquad\qquad\quad = 0$, *if* $x = 0$

is continuous at x = 0 but not differentiable at x = 0.

Solution : To show continuity, we have,

$$\lim_{x \to 0+} f(x) = \lim_{x \to 0+} x\sin\dfrac{1}{x}$$

$$= \lim_{x \to 0-} x\sin\dfrac{1}{x} = f(0) = 0$$

$\therefore \quad$ f(x) is continuous at x = 0.

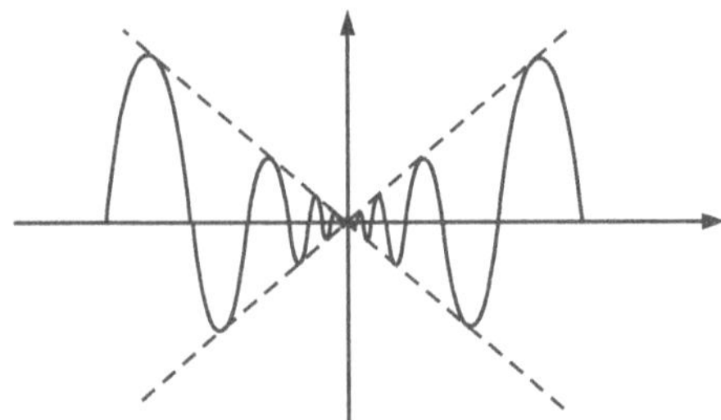

Fig. 1.6

Now, $\quad f'(0) = \displaystyle\lim_{h \to 0+} \frac{f(0 + h) - f(0)}{h} = \lim_{h \to 0} \frac{h \sin\frac{1}{h} - 0}{h}$

$$= \lim_{h \to 0} \sin\frac{1}{h},$$

does not exist

i.e. limit does not exists.   Therefore, $\;f'(0)$ does not exist.

Similarly, we can show that f' does not exist.

Hence $f'(0)$ does not exist. i.e. $x \sin \dfrac{1}{x}$ is not derivable at $x = 0$.

Example 1.4 : *If* $\qquad f(x) \;=\; x^4 \sin\dfrac{1}{x} \quad , \qquad x \neq 0$

$$= \; 0 \qquad , \qquad x = 0$$

show that this function is twice differentiable but not thrice.

Solution : We have, $\qquad f(x) \;=\; x^4 \sin\dfrac{1}{x}, \; x \neq 0$

$\therefore \qquad\qquad\qquad f'(x) \;=\; x^2\left[-\cos\dfrac{1}{x} + 4x \sin\dfrac{1}{x}\right]$

and $\qquad\qquad\quad f''(x) \;=\; -\sin\dfrac{1}{x} - 6x \cos\dfrac{1}{x} + 12x^2 \sin\dfrac{1}{x}$

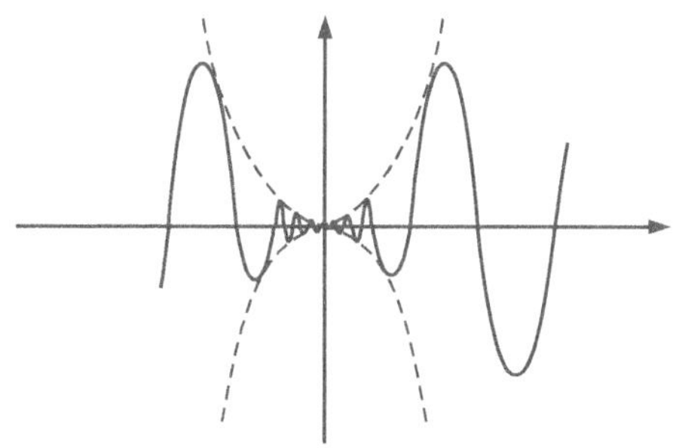

Fig. 1.7

Now, $\displaystyle\lim_{x \to 0+} f''(x) = \lim_{x \to 0+}\left\{-\sin\frac{1}{x} - 6x\cos\frac{1}{x} + 12x^2\sin\frac{1}{x}\right\}$

The limit does not exist.

Similarly, $\displaystyle\lim_{x \to 0-} f''(x)$ does not exist.

$\therefore \quad$ The function $f''(x)$ is not continuous at $x = 0$.

Hence $f'''(x)$ does not exist at $x = 0$ i.e. $x^4 \sin \dfrac{1}{x}$ is not thrice differentiable at $x = 0$.

Note : Similarly, we can show that the function,

$$f(x) = x^6 \sin \dfrac{1}{x} \ , \quad x \neq 0$$

$$= 0 \qquad , \quad x = 0$$

is differentiable three times but not at fourth time.

1.1.3 Algebra of Differential Functions

$\boxed{\textbf{Theorem 2}}$ If the functions f, g are derivable at c, then the functions.

(i) $kf, \ k \in \mathbb{R}$

(ii) $f + g$

(iii) $f \cdot g$

(iv) $f/g, \ g(c) \neq 0$ are differentiable at $x = c$.

Proof : (i)
$$\lim_{x \to c} \frac{(kf)\,x - (kf)\,c}{x - c} = \lim_{x \to c} \frac{k\,f(x) - k\,f(c)}{x - c}$$

$$= k \cdot \lim_{x \to c} \frac{f(x) - f(c)}{x - c}$$

$$= k\,f'(c)$$

(ii)
$$\lim_{x \to c} \frac{(f + g)\,x - (f + g)\,c}{x - c} = \lim_{x \to c} \frac{f(x) + g(x) - f(c) - g(c)}{x - c}$$

$$= \lim_{x \to c} \left\{ \frac{f(x) - f(c)}{x - c} + \frac{g(x) - g(c)}{x - c} \right\}$$

$$= \lim_{x \to c} \frac{f(x) - f(c)}{x - c} + \lim_{x \to c} \frac{g(x) - g(c)}{x - c} = f'(c) + g'(c)$$

Note : Similarly, we can prove that

$$\lim_{x \to c} \frac{(f - g)\,x - (f - g)\,c}{x - c} = f'(c) - g'(c)$$

(iii)
$$\lim_{x \to c} \frac{(fg)\,x - (fg)\,c}{x - c} = \lim_{x \to c} \frac{f(x)\,g(x) - f(c)\,g(c)}{x - c}$$

$$= \lim_{x \to c} \frac{f(x)\,g(x) - f(c)\,g(x) + f(c)\,g(x) - f(c)\,g(c)}{x - c}$$

$$=$$

$$\lim_{x \to c} \left\{ \frac{(f(x) - f(c)) \cdot g(x) + (g(x) - g(c))\,f(c)}{x - c} \right\}$$

$$= \lim_{x \to c} \frac{f(x) - f(c)}{x - c} \cdot \lim_{x \to c} g(x)$$

$$+ \lim_{x \to c} \frac{g(x) - g(c)}{x - c} \cdot \lim_{x \to c} f(c)$$

$$= f'(c) \, g(c) + g'(c) \, f(c)$$

(iv)
$$\lim_{x \to c} \frac{(f/g)\,x - (f/g)\,c}{x - c} = \lim_{x \to c} \frac{\left(\dfrac{f(x)}{g(x)} - \dfrac{f(c)}{g(c)}\right)}{x - c}$$

$$= \lim_{x \to c} \frac{f(x)\,g(c) - f(c)\,g(x)}{g(x)\,g(c)\,(x - c)}$$

$$= \lim_{x \to c} \frac{f(x)\,g(c) - f(c)\,g(c) + f(c)\,g(c) - f(c)\,g(x)}{g(x)\,g(c)\,(x - c)}$$

$$= \lim_{x \to c} \frac{1}{g(x)\,g(c)} \cdot \left[\frac{(f(x) - f(c))}{x - c} \cdot g(c) - \frac{(g(x) - g(c))}{x - c} \cdot f(c) \right]$$

$$= \frac{1}{g(c)\,g(c)} [f'(c) \cdot g(c) - g'(c) \cdot f(c)]$$

$$= \frac{f'(c)\,g(c) - g'(c)\,f(c)}{(g(c))^2} \text{ if } g(c) \neq 0.$$

Note : We have just now proved the result as

$$\left(\frac{f(x)}{g(x)}\right)'_{x = c} = \frac{f'(c)\,g(c) - g'(c)\,f(c)}{(g(c))^2}, \text{ if } g(c) \neq 0$$

Put
$$f(x) = 1 \text{ in above result}$$

$$\left(\frac{1}{g(x)}\right)'_{x = c} = \frac{0 \cdot g(c) - g'(c)\,f(c)}{(g(c))^2}$$

$$= -\frac{g'(c)}{(g(c))^2} \text{ (as } f(c) = 1)$$

$\boxed{\textbf{Theorem 3}}$ **(Caratheodary's Theorem) :** Suppose, I is an interval and $c \in I$. A function $f : I \to \mathbb{R}$ is differentiable at c if and only if there exists a function $g : I \to \mathbb{R}$, such that g is continuous at c and that satisfies

$$f(x) - f(c) = g(x)\,(x - c)$$

In this case, we have $g(c) = f'(c)$.

Proof : Assume that f is differentiable at c.

Then
$$f'(c) = \lim_{x \to c} \frac{f(x) - f(c)}{x - c} \text{ exists.}$$

Define a function,　$g(x) = \begin{cases} \dfrac{f(x) - f(c)}{x - c}, & \text{for } x \in I,\ x \neq c \\ f'(c), & \text{for } x = c \end{cases}$

Then g is continuous at c, since

$$\lim_{x \to c} g(x) = \lim_{x \to c} \frac{f(x) - f(c)}{x - c} = f'(c) = g(c).$$

Further, if $x = c$, then $g(x)(x - c) = 0 = f(x) - f(c)$,

while　if $x \neq c$, then $f(x) - f(c) = g(x)(x - c)$, $\forall\, x \in I$

Thus, g is the required function.

Conversely, suppose there exists a function $g : I \to \mathbb{R}$, such that g is continuous at c and $f(x) - f(c) = (x - c)\, g(x)$, $\forall\, x \in I$.

Then for $x \neq c$, we have, $g(x) = \dfrac{f(x) - f(c)}{x - c}$.

$\therefore$ 　　　　$\lim_{x \to c} g(x) = \lim_{x \to c} \dfrac{f(x) - f(c)}{x - c}$

$\Rightarrow\ g(c) = f'(c) \Rightarrow f$ is differentiable at c and $f'(c) = g(c)$.　　■

Chain rule : Chain rule gives a formula for finding the derivative of a composite function, gof in terms of the derivatives of g and f.

$\boxed{\textbf{Theorem 4}}$ **(chain rule) :** Let $g : I_1 \to R$ and $f : I_2 \to R$ be functions such that $f(I_2) \subseteq I_1$ and $x_0 \in I_2$. If f is derivable at x_0 and g is derivable at $f(x_0)$, then the composite function gof is derivable at x_0 and

$$(gof)'(x_0) = g'(f(x_0)) \cdot f'(x_0)$$

Derivative of a inverse function :

We know that if f is continuous strictly monotonic function on an interval I_1, then its inverse function $g = f^{-1}$ is defined on the interval $I_2 = f(I_1)$ and satisfies the relation.

$$g(f(x)) = x \quad \text{for } x \in I_1 \qquad \ldots (1)$$

If $x_0 \in I_1$ and $y_0 = f(x_0)$ and if $f'(x_0)$ and $g'(y_0)$ both exist, then differentiating equation (1) by chain rule, we have

$$g'(f(x_0)) \cdot f'(x_0) = 1$$

Thus, if $f'(x_0) \neq 0$, we get

$$g'(y_0) = \frac{1}{f'(x_0)}$$

Remark 1 : In proof of theorem 2, whenever we have distributed limits over sum, product provided these limits exist.

2 : Function f is differentiable at $x = c$, iff graph of f is smooth at point $(c, f(c))$.

Illustrative Examples

Example 1.5 : *Let $f : \mathbb{R} \to \mathbb{R}$ be defined by $f(x) = x^3 + 7x + 1$. Find the derivative of f^{-1} at 9.*

Solution : Note here that f is monotonic increasing and

$$f'(x) = 3x^2 + 7 \neq 0, \ \forall \ x \in \mathbb{R}.$$

Therefore the function f^{-1} is differentiable and

$$\frac{d}{dy} f^{-1}(y) = \frac{1}{f'(x)}, \text{ where } f(x) = y$$

Note that, $f(1) = 9$, and $f'(x) = 3x^2 + 7$.

$$\therefore \qquad \frac{d}{dy} f^{-1}(9) = \frac{1}{f'(1)}$$

$$= \frac{1}{3+7} = \frac{1}{10}$$

Hence, the derivative of f^{-1} at 9 is $\frac{1}{10}$.

Example 1.6 : *Let $f : [-1, 1] \to \mathbb{R}$ be defined by $f(x) = \sin^{-1} x$.*

Find $f'(x)$.

Solution : Let $g : \left[\dfrac{-\pi}{2}, \dfrac{\pi}{2}\right] \to \mathbb{R}$ be the function defined by $g(x) = \sin x$.

Then g is monotonic increasing and differentiable with

$$g'(x) \neq 0 \ \forall \ x \in \left(\frac{-\pi}{2}, \frac{\pi}{2}\right).$$

$$\therefore \qquad \frac{d}{dy} f(y) = \frac{1}{g'(x)}, \text{ where, } y = g(x) = \sin x$$

Here,
$$g'(x) = \cos x = \sqrt{1 - \sin^2 x} = \sqrt{1 - y^2}$$

$$\therefore \quad f'(y) = \frac{1}{\sqrt{1 - y^2}}$$

Example 1.7 : *Let $f(x) = \tan^{-1} x$, $x \in \mathbb{R}$. Find $f'(x)$.*

Solution : Let $g : \left[\dfrac{-\pi}{2}, \dfrac{\pi}{2}\right] \to \mathbb{R}$ be a function defined by

$g(x) = \tan x$. Then g is monotonic increasing and

$$g'(x) = \sec^2 x \neq 0 \ \forall \ x \in \left(\frac{-\pi}{2}, \frac{\pi}{2}\right)$$

$$\therefore \quad \frac{d}{dy} g^{-1}(y) = \frac{d}{dy} f(y) = \frac{1}{g'(x)}, \text{ where } y = g(x)$$

$$\therefore \quad \frac{d}{dy} f(y) = f'(y) = \frac{1}{\sec^2 x} = \frac{1}{1 + \tan^2 x} = \frac{1}{1 + y^2}$$

$$\text{Thus,} \quad \frac{d}{dy} (\tan^{-1} y) = \frac{1}{1 + y^2}.$$

Example 1.8 : Let $f(x) = |x^2 - 1|$, $x \in \mathbb{R}$. Discuss the differentiability of f on $\mathbb{R}$.

Solution :

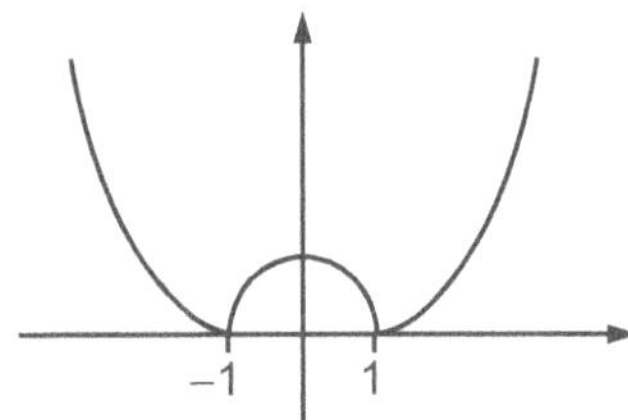

Fig. 1.8 : Graph of f

Note that,
$$f(x) = \begin{cases} x^2 - 1 & ; \text{ if } x^2 > 1 \\ 1 - x^2 & ; \text{ if } |x| \leq 1 \end{cases}$$

$$\therefore \quad f'_+(1) = \lim_{x \to 1^+} \frac{f(x) - f(1)}{x - 1} = \lim_{x \to 1^+} \frac{x^2 - 1 - 0}{x - 1}$$

$$= \lim_{x \to 1^+} \frac{(x - 1)(x + 1)}{x - 1} = 2,$$

and
$$f'_-(1) = \lim_{x \to 1^-} \frac{f(x) - f(1)}{x - 1}$$

$$= \lim_{x \to 1^-} \frac{1 - x^2 - 0}{x - 1} = -2$$

$\therefore \qquad f'_+(1) \neq f'_-(1) \Rightarrow f'(1)$ does not exists.

Hence, f is not differentiable at 1. Similarly, f is not differentiable at $x = -1$.

Now, if $c \in (-1, 1)$, then

$$f'(x) = \lim_{x \to c} \frac{f(x) - f(c)}{x - c} = \lim_{x \to c} \frac{1 - x^2 - 1 + c^2}{x - c} = -2c$$

and if $|c| > 1 \Rightarrow c \in (-\infty, -1) \cup (1, \infty)$. Then $f'(c) = 2c$.

Thus, f is differentiable everywhere on $\mathbb{R}$, except at 1 and -1.

1.2 The Mean Value Theorems

The mean value theorem gives the relation between the values of the function and values of its derivative. In this article we shall see the mean value theorem and its applications.

Theorem 5 (Rolle's Theorem) - (Oct. 2011; April 2012, 2013) : Suppose that a function f is (i) continuous on the closed and bounded interval [a, b], (ii) derivable on the open interval (a, b) and (iii) f(a) = f(b), then there exists at least one point $c \in (a, b)$ such that $f'(c) = 0$.

Proof : Given that, f is continuous on [a, b]. Therefore, f is bounded and attains its bounds at least once in [a, b]. Let M and m be respectively the supremum and infimum of f(x) in [a, b]. Clearly that $m \leq M$.

Case 1 : Suppose M = m, then f is a constant function on [a, b]. Consequently at any point $c \in (a, b)$, we should have $f'(c) = 0$. (by definition of derivative)

Case 2 : Suppose m < M. In this case, at least one of the bounds M and m is different from f(a) = f(b). For the sake of definiteness, suppose $M \neq f(a)$. Let $M = f(c) \neq f(a)$, $c \in (a, b)$. Then $f'(c)$ exists.

$$\therefore \qquad f'(c) = f'_+(c) = f'_-(c)$$

We have, $\qquad f(x) \le f(c) = M \quad \forall\, x \in [a, b]$

$\therefore \qquad \dfrac{f(x) - f(c)}{x - c} \ge 0, \ \text{if} \ \ x < c$

and $\qquad \dfrac{f(x) - f(c)}{x - c} \le 0, \ \text{if} \ \ x > c$

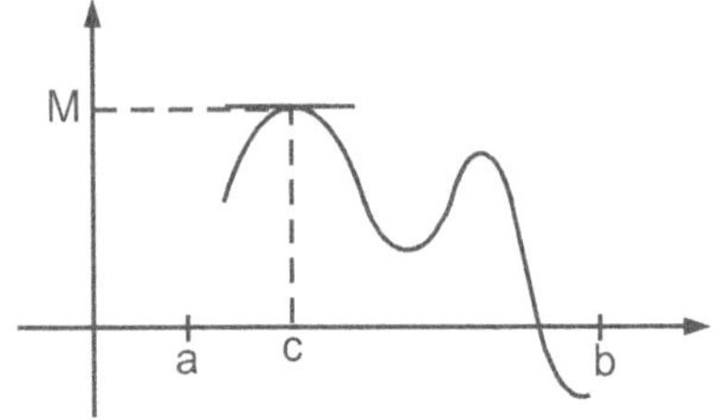

Fig. 1.9

$\therefore \qquad f'_{-}(c) = \lim_{x \to c^{-}} \dfrac{f(x) - f(c)}{x - c} \ge 0 \qquad \ldots \text{(i)}$

and $\qquad f'_{+}(c) = \lim_{x \to c^{+}} \dfrac{f(x) - f(c)}{x - c} \le 0 \qquad \ldots \text{(ii)}$

From equation (i) and (ii), we have $f'(c) = 0$.

Similar argument can be given if $m \ne f(a)$. Hence, the theorem is proved. ■

Remarks :

1. Geometrically Rolle's theorem states that under the given conditions there exists at least one point within the interval (a, b) such that the tangent at the point is parallel to the X-axis (See Fig. 1.9 and 1.10).

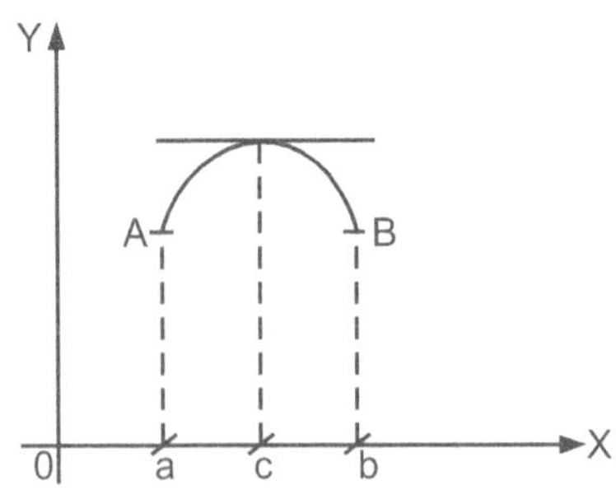

Fig. 1.10

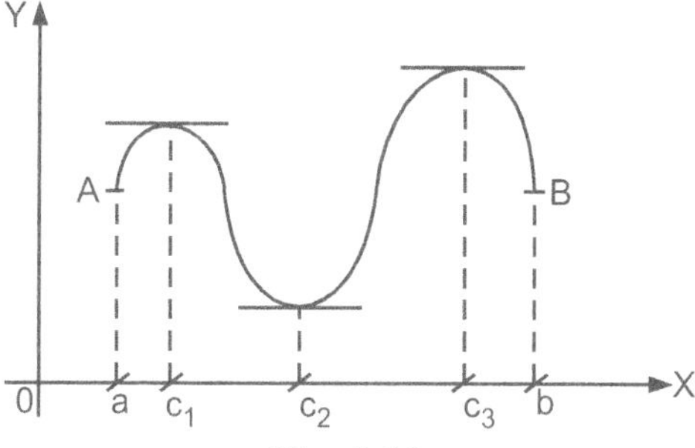

Fig. 1.11

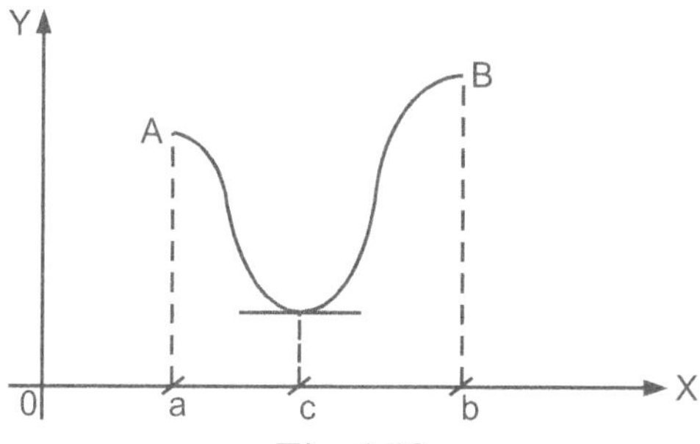

Fig. 1.12

2. Let $f(x)$ be a polynomial function defined on an interval $[a, b]$. Then $f(x)$ is continuous and derivable on $[a, b]$. If $f(a) = 0$ and $f(b) = 0$ i.e. a and b are the roots of the polynomial function $f(x)$ i.e. the third condition of Rolle's theorem is satisfied, then by Rolle's theorem these exists at least one point $c \in (a, b)$ such that $f'(c) = 0$. *In another words between any two roots of $f(x) = 0$, there exists at least one root of the equation $f'(x) = 0$.*

3. Rolle's theorem states that if f has its maximum (or minimum) value at an interior point $c \in (a, b)$, then either the derivative of f at c is equal to zero or derivative of f at c does not exists. For example, consider the function $f(x) = |x|$ on $[-1, 1]$. The function f has an interior minimum at $x = 0$. But the derivative of f at $x = 0$ does not exists.

4. The third condition in the Rolle's theorem viz $f(a) = f(b)$ is sufficient but not necessary for the conclusion of the theorem. It is clear from the Fig. 1.11.

As a consequence of Rolle's theorem, we obtain the following fundamental mean value theorem.

Theorem 6 (**Lagrange's Mean Value Theorem**) :

Suppose that a function f is continuous on closed and bounded interval [a, b] and derivable on an open interval (a, b). Then there exists at least one point $c \in (a, b)$, such that

$$f'(c) \; = \; \frac{f(b) - f(a)}{b - a}$$

Proof : Consider a function ϕ defined on [a, b] as $\phi(x) = f(x) - Ax$, where the constant A is to be chosen such that $\phi(a) \; = \; \phi(b)$.

So that, $\phi(a) \; = \; f(a) - Aa$

and $\phi(b) \; = \; f(b) - Ab$

Thus, $\phi(a) = \phi(b) \Rightarrow f(a) - Aa \; = \; f(b) - Ab$

$$\Rightarrow A = \frac{f(b) - f(a)}{b - a} \qquad \qquad \text{... (i)}$$

Function $\phi(x)$ satisfies all the conditions of Rolle's theorem. Hence, by Rolle's theorem there exists at least one point $c \in (a, b)$ such that $\phi'(c) \; = \; 0$.

But $\phi'(x) \; = \; f'(x) - A$

$\therefore$ $\phi'(c) \; = \; f'(c) \; - A$

and $\phi'(c) \; = \; 0 \Rightarrow f'(c) - A \; = \; 0$

$$\Rightarrow A \; = \; f'(c) \qquad \qquad \text{... (ii)}$$

From (i) and (ii), we have

$$f'(c) \; = \; \frac{f(b) - f(a)}{b - a}$$

Hence, the theorem is proved.   ■

If we write $b - a \; = \; h$ in above theorem, then $b \; = a + h$.

so that $c \in (a, a + h) \Rightarrow a < c < a + h$

$$\Rightarrow 0 < c - a < h$$

$$\Rightarrow 0 < \frac{c - a}{h} < 1$$

Let $\theta \; = \; \dfrac{c - a}{h}$ $\therefore \; 0 < \theta < 1$

$\Rightarrow$ We can write $c = a + \theta h$ with $0 < \theta < 1$.

$\therefore$ $f'(a + \theta h) \; = \; \dfrac{f(b) - f(a)}{b - a}$ for some $0 < \theta < 1$

Hence, we have the following :

Another form of Mean Value Theorem :

Suppose that a function f is (i) continuous on the closed interval [a, a + h] and (ii) derivable on the open interval (a, a + h). Then there exists at least one point $c = a + \theta h$, where $0 < \theta < 1$, such that $f(a + h) = f(a) + h\,f'(a + \theta h)$.

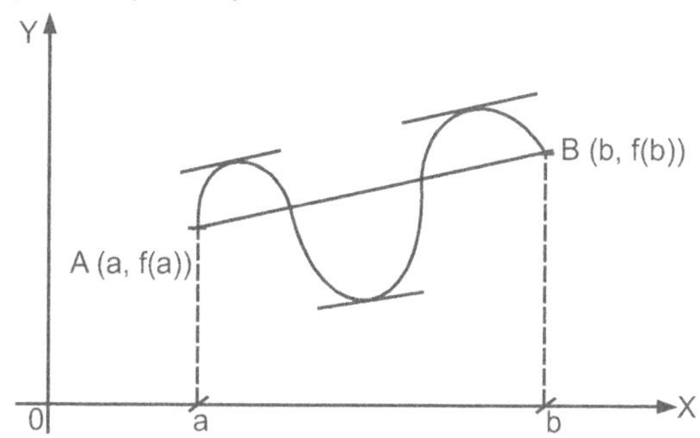

Fig. 1.13

Remark :

1. Rolle's theorem is a special case of Lagrange's Mean value theorem when $f(a) = f(b)$.
2. Geometrically, the Lagrange's mean value theorem states that there is at least one point c on the curve $y = f(x)$ at which the tangent is parallel to the line segment through the points (a, f(a)) and (b, f(b)). See Fig. 1.13.

Now, we shall give some important consequences of the Mean Value Theorem. ■

Theorem 7 Suppose that a function f is continuous on the closed and bounded interval [a, b] and derivable on an open interval (a, b) with $f'(x) = 0$ for every $x \in (a, b)$. Then f is constant on [a, b].

Proof : Let $x \in (a, b) \Rightarrow x > a$.

Consider the closed interval [a, x]. The function f is obviously continuous on the closed interval [a, x] and derivable on the open interval (a, x). Therefore, by Lagrange's mean value theorem, there exists a point $c \in (a, x)$ such that.

$$\frac{f(x) - f(a)}{x - a} = f'(c)$$

i.e. $\qquad\qquad f(x) - f(a) = (x - a)\,f'(c)$

But $\qquad\qquad f'(c) = 0 \Rightarrow f(x) - f(a) = 0$

$$\Rightarrow f(x) = f(a) \text{ for every } x \in (a, b)$$

i.e. f is constant on [a, b]. ■

Theorem 8 Suppose that the functions f and g are continuous on [a, b]

and derivable on (a, b) with $f'(x) = g'(x)$, $\forall\, x \in (a, b)$.

Then $f(x) = g(x) + c$ for some constant c.

Proof : Let $\qquad\qquad\qquad\qquad \phi(x) = f(x) - g(x),\ x \in [a, b]$

Now, we shall show that $\phi(x)$ is constant on [a, b].

We have, $\qquad\qquad\qquad\qquad \phi'(x) = f'(x) - g'(x)$

$$= 0 \quad \forall\, x \in (a, b)$$

$\therefore \qquad\qquad\qquad\qquad \phi'(x) = 0, \quad \forall\, x \in (a, b)$

Hence, by theorem (7) $\phi(x)$ is a constant function on [a, b].

Therefore $\qquad\qquad\qquad \phi(x) = c \qquad\qquad \forall\, x \in [a, b]$

$\Rightarrow \qquad\qquad\qquad f(x) - g(x) = c, \qquad\qquad \forall\, x \in [a, b]$

$\Rightarrow \qquad\qquad\qquad f(x) = g(x) + c\ \ \forall\, x \in [a, b]$

Increasing and Decreasing Functions :

The function f is said to be **increasing** on the interval [a, b] if $x_1 < x_2 \Rightarrow f(x_1) \le f(x_2)$ and f is said to be **decreasing** on the interval [a, b] if $x_1 < x_2 \Rightarrow f(x_1) \ge f(x_2)$.

Similarly, the function f is said to be **strictly increasing** on the interval [a, b], if $x_1 < x_2 \Rightarrow f(x_1) < f(x_2)$ and f is said to be **strictly decreasing** on the interval [a, b], if $x_1 < x_2 \Rightarrow f(x_1) > f(x_2)$.

A function is said to be **monotonic** if it is either increasing or decreasing.

Theorem 9 If f is continuous on [a, b] and f is derivable on (a, b) with

$f'(x) > 0$, $\forall\, x \in (a, b)$, then $f(x)$ is strictly increasing function on [a, b].

Proof : Let $a \le x_1 < x_2 \le b$.

Obviously the function is continuous on closed interval $[x_1, x_2]$ derivable on the open interval (x_1, x_2). Then by mean value theorem there exists a point $c \in (x_1, x_2)$ such that

$$\frac{f(x_2) - f(x_1)}{x_2 - x_1} = f'(c)$$

$\Rightarrow \qquad\qquad\qquad \dfrac{f(x_2) - f(x_1)}{x_2 - x_1} > 0 \qquad\qquad\qquad (\because\ f'(c) > 0)$

$\Rightarrow \qquad\qquad\qquad f(x_2) - f(x_1) > 0 \qquad\qquad\qquad (\because\ x_2 - x_1 > 0)$

$\Rightarrow \qquad\qquad\qquad f(x_2) > f(x_1), \quad \text{where } x_2 > x_1.$

Thus, f is strictly increasing on [a, b].

$\boxed{\textbf{Theorem 10}}$ If f is continuous on [a, b] and f is derivable on (a, b) with $f'(x) < 0$, $\forall\, x \in$ (a, b), then f(x) is strictly decreasing function on [a, b].

Proof : Left to the reader. ∎

$\boxed{\textbf{Theorem 11}}$ **(Cauchy's Mean Value Theorem) :** Suppose that the function f and g are (i) continuous on [a, b], (ii) derivable on (a, b) and (iii) $g'(x) \neq 0$, $\forall\, x \in$ (a, b). Then there exists a point $c \in$ (a, b), such that

$$\frac{f(b) - f(a)}{g(b) - g(a)} = \frac{f'(c)}{g'(c)}$$

Proof : We claim that $g(b) - g(a) \neq 0$. If $g(b) = g(a)$, then the function g satisfies all the conditions of Rolle's theorem. Hence, there exists at least one point $c \in$ (a, b), such that $g'(c) = 0$. This is contrary to the condition (ii) of the theorem. Hence, $g(b) - g(a) \neq 0$.

Consider the function ϕ defined on [a, b] as $\phi(x) = f(x) - A\, g(x)$, where the constant A is to be chosen such that

$$\phi(a) = \phi(b)$$

So that,

$$\phi(a) = f(a) - Ag(a) \quad \text{and} \quad \phi(b) = f(b) - A\, g(b)$$

$$\therefore \qquad \phi(a) = \phi(b) \Rightarrow f(a) - A\, g(a) = f(b) - A\, g(b)$$

$$\Rightarrow A = \frac{f(b) - f(a)}{g(b) - g(a)} \qquad \qquad \text{... (i)}$$

The function $\phi(x)$ satisfies all the conditions of Rolle's theorem. By Rolle's theorem, there exists at least one point $c \in$ (a, b), such that $\phi'(c) = 0$.

But

$$\phi'(x) = f'(x) - A\, g'(x)$$

$$\therefore \qquad \phi'(c) = 0 \Rightarrow f'(c) - A\, g'(c) = 0$$

$$\Rightarrow A = \frac{f'(c)}{g'(c)} \qquad \qquad \text{... (ii)}$$

From (i) and (ii), we have

$$\frac{f'(c)}{g'(c)} = \frac{f(b) - f(a)}{g(b) - g(a)}$$

Hence, the theorem. ∎

Another form of Cauchy's Mean Value Theorem :

If the functions f and g are (i) continuous on [a, a + h], (ii) derivable on (a, a + h) and $g'(x) \neq 0$, $\forall\, x \in$ (a, a + h), then there exists at least one point $a + \theta h \in$ (a, a + h), where $0 < \theta < 1$, such that

$$\frac{f(a+h) - f(a)}{g(a+h) - g(a)} = \frac{f'(a + \theta h)}{g'(a + \theta h)}$$

Note : In Cauchy's mean value theorem if we put $g(x) = x$, Cauchy's mean value theorem reduces to Lagrange's mean value theorem.

| Theorem 12 | **(Bernoulli's inequality) :** If $r > 1$, then

$(1 + x)^r \geq 1 + rx$, for all $x > -1$. Equality if and only if $x = 0$.

Proof : We now prove Bernoulli's inequality for any $r \in \mathbb{Q}$, using mean value theorem.

(i) If $x > 0$.

We apply mean value theorem to $f(x) = (1 + x)^r$ in the interval $[0, x]$.

There exists $c \in (0, x)$ such that

$$f(x) - f(0) = f'(c)(x - 0)$$

$$\therefore \qquad (1 + x)^r - 1 = r(1 + c)^{r-1} \cdot x$$

Since $c > 0$ and $r - 1 > 0 \Rightarrow (1 + c)^{r-1} > 1$.

Hence, $\qquad (1 + x)^r > 1 + rx$

(ii) If $-1 < x < 0$, we apply mean value theorem to $f(x) = (1 + x)^r$ in the interval $[x, 0]$.

There exists $c \in (x, 0)$ such that

$$f(0) - f(x) = f'(c)(0 - x)$$

$$\therefore \qquad 1 - (1 + x)^r = r(1 + c)^{r-1}(-x)$$

i.e. $\qquad (1 + x)^r - 1 = r(1 + c)^{r-1} \cdot x$

Since $-1 < c < 0$ and $r - 1 > 0$,

$$\Rightarrow \qquad 0 < (1 + c) < 1$$

$$\Rightarrow \qquad 0 < (1 + c)^{r-1} < 1$$

$$\therefore \qquad r(1 + c)^{r-1} x > rx \qquad\qquad (\because x < 0)$$

Hence, $\qquad (1 + x)^r > 1 + rx$

(iii) If $x = 0$. The equality holds. ∎

Illustrative Examples

Example 1.9 : *Verify Rolle's theorem for the function*
$$f(x) = 2x^3 + x^2 - 4x - 2 \text{ on } [-\sqrt{2}, \sqrt{2}].$$

Solution : We have $\qquad f(x) = 2x^3 + x^2 - 4x - 2$

$$= (x^2 - 2)(2x + 1)$$

$$\therefore \qquad f(-\sqrt{2}) = 0 = f(\sqrt{2})$$

$f(x)$ is a polynomial function, so it is continuous on $[-\sqrt{2}, \sqrt{2}]$ and differentiable on $(-\sqrt{2}, \sqrt{2})$. All the conditions of Rolle's theorem are satisfied. Therefore, the derivative of $f(x)$ must vanish for at least one value of $x \in (-\sqrt{2}, \sqrt{2})$.

Now,
$$f'(x) = 6x^2 + 2x - 4$$
$$f'(x) = 0 \Rightarrow 6x^2 + 2x - 4 = 0$$
$$\Rightarrow (3x - 2)(x + 1) = 0$$
$$\Rightarrow x = -1 \quad \text{or} \quad 2/3$$

Since both the points -1, $2/3$ lie in the open interval $(-\sqrt{2}, \sqrt{2})$, Rolle's theorem is verified.

Example 1.10 : *Verify Rolle's theorem for the function*

$$f(x) = e^x (\sin x - \cos x) \text{ on } \left[\frac{\pi}{4}, \frac{5\pi}{4}\right].$$

Solution : We have, $\quad f(x) = e^x (\sin x - \cos x)$

$$\therefore \quad f(\pi/4) = e^{\pi/4}\left(\sin\frac{\pi}{4} - \cos\frac{\pi}{4}\right)$$

$$= e^{\pi/4}\left(\frac{1}{\sqrt{2}} - \frac{1}{\sqrt{2}}\right) = 0$$

and
$$f(5\pi/4) = e^{5\pi/4}\left(\sin\frac{5\pi}{4} - \cos\frac{5\pi}{4}\right)$$

$$= e^{5\pi/4}\left(-\frac{1}{\sqrt{2}} + \frac{1}{\sqrt{2}}\right) = 0$$

$$\therefore \quad f(\pi/4) = 0 = f(5\pi/4)$$

The function $f(x)$ is continuous on $[\pi/4, 5\pi/4]$ and differentiable on $(\pi/4, 5\pi/4)$.

All the conditions of Rolle's theorem are satisfied. Therefore, the derivative of $f(x)$ must vanish for at least one value of $x \in (\pi/4, 5\pi/4)$.

We have, $\quad f'(x) = e^x (\cos x + \sin x) + e^x (\sin x - \cos x)$

$$= 2e^x \sin x$$

$$\therefore \quad f'(x) = 0 \Rightarrow 2e^x \sin x = 0$$

$$\Rightarrow \sin x = 0$$

$$\Rightarrow x = 0, \pm \pi, \pm 2\pi, \ldots$$

Out of these values, only $x = \pi$ lies in the interval $(\pi/4, 5\pi/4)$.

Example 1.11 : *If $f(x) = (x - 2) \log x$, show that the equation $x \log x = 2 - x$ is satisfied by at least one value of x lying between 1 and 2.*

Solution : We have, $\qquad f(x) = (x - 2) \log x$

$\therefore \qquad\qquad\qquad f(1) = 0 = f(2)$

The function $f(x)$ is continuous on $[1, 2]$ and differentiable on $(1, 2)$. All the conditions of Rolle's theorem are satisfied.

$\therefore$ Derivative of $f(x)$ must vanish for at least one value of $x \in (1, 2)$.

Here, $\qquad\qquad f'(x) = \dfrac{x - 2}{x} + \log x = \dfrac{(x - 2) + x \log x}{x}$

$\therefore \qquad\qquad f'(x) = 0 \Rightarrow (x - 2) + x \log x = 0$

By Rolle's theorem, between any two roots viz. $x = 1$ and $x = 2$ of $f(x)$, there exists at least one root of the equation $f'(x) = 0$, hence $x \log x = 2 - x$ is satisfied by at least one value of x lying between 1 and 2.

Example 1.12 : *Verify the mean value theorem for $f(x) = \log x$ on $[1, e]$.*

Solution : We have, $\qquad f(x) = \log x$

The function $f(x)$ is continuous on $[1, e]$ and differentiable on $(1, e)$. All the conditions of the mean value theorem are satisfied. Hence, by Lagrange's mean value theorem, we have

$$\frac{f(e) - f(1)}{e - 1} = f'(c)$$

$$\Rightarrow \qquad \frac{\log e - \log 1}{e - 1} = \frac{1}{c} \Rightarrow \frac{1 - 0}{e - 1} = \frac{1}{c}$$

$\therefore \quad c = e - 1$ which lies in $(1, e)$.

Example 1.13 : *Verify Cauchy's mean value theorem for the functions $f(x) = \dfrac{1}{x^2}$ and $g(x) = \dfrac{1}{x}$ in $[a, b]$, $a > 0$. Show that the point c is harmonic mean of a and b.*

Solution : Here f and g satisfies all conditions of C.M.V.T.

$\therefore \qquad$ By C.M.V.T. $\quad \dfrac{f'(c)}{g'(c)} = \dfrac{f(b) - f(a)}{g(b) - g(a)}$ for some $c \in (a, b)$.

$$\Rightarrow \qquad \frac{\left(-\dfrac{2}{c^3}\right)}{\left(-\dfrac{1}{c^2}\right)} = \frac{\dfrac{1}{b^2} - \dfrac{1}{a^2}}{\dfrac{1}{b} - \dfrac{1}{a}}$$

$$\Rightarrow \qquad \frac{2}{c} = \frac{a^2 - b^2}{ab\,(a-b)} = \frac{(a-b)\,(a+b)}{ab\,(a-b)} = \frac{a+b}{ab}$$

$$\therefore \qquad c = \frac{2ab}{a+b}$$

which is harmonic mean of a and b.

Example 1.14 : *If in the Cauchy's mean value theorem, we write* $f(x) = e^x$ *and* $g(x) = e^{-x}$ *show that 'c' is the arithmetic mean of a and b.*

Solution : Here f and g satisfies all conditions of C.M.V.T.

We have,
$$\frac{f(b) - f(a)}{g(b) - g(a)} = \frac{e^b - e^a}{e^{-b} - e^{-a}} = -e^{a+b}$$

and
$$\frac{f'(x)}{g'(x)} = \frac{e^x}{-e^{-x}} \qquad \therefore \frac{f'(c)}{g'(c)} = \frac{e^c}{-e^{-c}} = -e^{2c}$$

hence,
$$-e^{2c} = -e^{a+b}$$

$$\Rightarrow \qquad 2c = a+b \qquad \therefore \quad c = \frac{a+b}{2}$$

i.e. c is the A.M. of a and b.

Example 1.15 : *Under suitable conditions, prove that in the interval* (a, b) *there exists at least one point c, such that*

$$\begin{vmatrix} f(a) & f(b) \\ \phi(a) & \phi(b) \end{vmatrix} = (b-a) \begin{vmatrix} f(a) & f'(c) \\ \phi(a) & \phi'(c) \end{vmatrix}$$

Solution : Define function

$$F(x) = \left(\frac{x-a}{b-a}\right) \begin{vmatrix} f(a) & f(b) \\ \phi(a) & \phi(b) \end{vmatrix} - \begin{vmatrix} f(a) & f(x) \\ \phi(a) & \phi(x) \end{vmatrix} \text{ on } [a, b]$$

Since f, ϕ are continuous on [a, b], F is continuous on [a, b].

Since f, ϕ are differentiable on (a, b), F is differentiable on (a, b)

Also,
$$F(a) = F(b) = 0$$

$\therefore$ By Roll's theorem $\exists\, c \in$ (a, b) s. t. $F'(c) = 0$

i.e.
$$\begin{vmatrix} f(a) & f(b) \\ \phi(a) & \phi(b) \end{vmatrix} = (b-a) \begin{vmatrix} f(a) & f'(c) \\ \phi(a) & \phi'(c) \end{vmatrix}$$

Example 1.16 : *Separate the intervals in which the polynomial* $x^3 + 8x^2 + 5x - 2$ *is increasing or decreasing.*

Solution : We have $f(x) = x^3 + 8x^2 + 5x - 2$

$$f'(x) = 3x^2 + 16x + 5 = (3x + 1)\,(x + 5)$$

$$f'(x) = 0 \Rightarrow x = -1/3 \quad \text{or} \quad x = -5$$

So, we consider three intervals $(-\infty, -5)\left(-5, -\frac{1}{3}\right)$ and $\left(-\frac{1}{3}, \infty\right)$ and

see in which interval the value of $f'(x)$ is positive or negative.

(i)   Consider the interval $(-\infty, -5)$. In this interval

       $(3x + 1) < 0$ and $(x + 5) < 0$

$\therefore$ $f'(x) = (3x + 1)(x + 5) > 0$

$\Rightarrow$ $f(x)$ is increasing on $(-\infty, -5)$.

(ii)   Consider the interval $(-5, -1/3)$.

       In this interval $(3x + 1) < 0$ and $(x + 5) > 0$

$\therefore$ $f'(x) = (3x + 1)(x + 5) < 0$

$\Rightarrow$ $f(x)$ is decreasing on $(-5, -1/3)$.

(iii) Consider the interval $(-1/3, \infty)$.

       In this interval $(3x + 1) > 0$ and $(x + 5) > 0$

$\therefore$ $f'(x) = (3x + 1)(x + 5) > 0$

$\Rightarrow$ $f(x)$ is increasing on $(-1/3, \infty)$.

Example 1.17 : *Show that*

$$\forall x > 0, \quad x - \frac{x^2}{2} < \log(1 + x) < x - \frac{x^2}{2(1 + x)}.$$

Solution : Consider the function

$$f(x) = \log(1 + x) - \left(x - \frac{x^2}{2}\right)$$

$\therefore$ $$f'(x) = \frac{1}{1 + x} - 1 + x = \frac{x^2}{1 + x} > 0, \quad \forall x > 0$$

except at $x = 0$

and $$f(x) = 0 \quad \text{at} \quad x = 0$$

$\therefore$ $f(x)$ is increasing in $[0, \infty)$ and $f(x) > f(0)$ $\forall x > 0$.

hence $$\log(1 + x) > x - \frac{x^2}{2}, \quad \forall x > 0$$

Next consider the function

$$f(x) = x - \frac{x^2}{2(1 + x)} - \log(1 + x)$$

$\therefore$ $$f'(x) = 1 - \frac{2(1 + x) \cdot 2x - 2x^2}{4(1 + x)^2} - \frac{1}{1 + x}$$

$$= \frac{x^2}{2(1 + x)^2} > 0, \quad \forall x > 0 \quad \text{except at } x = 0$$

Again $f(x) = 0$ at $x = 0$.

$\therefore$ $f(x)$ is increasing in $[0, \infty)$ and $f(x) > f(0)$ $\forall\, x > 0$

i.e. $x - \dfrac{x^2}{2\,(1+x)} > \log\,(1+x),\ \forall\, x > 0$

Hence, $x - \dfrac{x^2}{2} < \log\,(1+x) < x - \dfrac{x^2}{2\,(1+x)}$

Example 1.18 : *Prove that $\dfrac{x}{1+x^2} < tan^{-1}\, x < x, \quad \forall x > 0.$*

Solution : Let $f(x) = \tan^{-1} x - \dfrac{x}{1+x^2}$

$\therefore$ $f'(x) = \dfrac{1}{1+x^2} - \left[\dfrac{(1+x^2) - 2x^2}{(1+x^2)^2}\right]$

$\qquad\qquad\quad = \dfrac{1}{1+x^2} - \dfrac{1-x^2}{(1+x^2)^2} = \dfrac{2x^2}{(1+x^2)^2} > 0$

$\Rightarrow$ f is monotonic increasing.

By theorem $f(x) > f(0)$.

$\Rightarrow$ $\tan^{-1} x - \dfrac{x}{1+x^2} > 0$

$\Rightarrow$ $\dfrac{x}{1+x^2} < \tan^{-1} x$ ... (i)

Next let $g(x) = x - \tan^{-1} x$

$\therefore$ $g'(x) = 1 - \dfrac{1}{1+x^2} = \dfrac{x^2}{1+x^2} > 0$

$\Rightarrow$ $g(x)$ is monotonic increasing.

$\Rightarrow$ $g(x) > g(0)$

$\Rightarrow$ $g(x) > 0$

$\Rightarrow$ $x - \tan^{-1} x > 0$

$\Rightarrow$ $\tan^{-1} x < x$

(i) and (ii) $\Rightarrow \dfrac{x}{1+x^2} < \tan^{-1} x < x, \quad \forall\, x > 0.$

Example 1.19 : *Prove that $e^x \geq 1 + x,\ for\ x \geq 0$ and equality holds iff $x = 0$.*

Solution :   Let $f(x) = e^x$ $\therefore$ $f'(x) = e^x > 1,\ \forall\, x > 0.$

(i) If $x = 0$, it is obvious that equality holds.

(ii) If $x > 0$, we apply Mean Value Theorem to the function in the interval $[0, x]$, then for some c with $0 < c < x$,

we have,
$$\frac{e^x - e^0}{x - 0} = e^c$$

i.e.
$$e^x - 1 = x\, e^c$$

But
$$e^c > 1 \quad \therefore \quad e^x - 1 > x$$

$\Rightarrow$
$$e^x > 1 + x, \quad \text{for } x > 0$$

Example 1.20 : *If $a < 1$, $b < 1$ and $b > a$, then prove that*

$$\frac{b - a}{\sqrt{1 - a^2}} < \sin^{-1} b - \sin^{-1} a < \frac{b - a}{\sqrt{1 - b^2}}$$

Solution : Let $\qquad f(x) = \sin^{-1} x, \quad a < x < b$

$\therefore \qquad\qquad f'(x) = \dfrac{1}{\sqrt{1 - x^2}}$

By Mean value theorem.

$$\frac{\sin^{-1} b - \sin^{-1} a}{b - a} = \frac{1}{\sqrt{1 - c^2}} \quad \text{for some } c \in (a, b)$$

Now, $\qquad\qquad\qquad a < c < b$

$$a^2 < c^2 < b^2$$

$$1 - a^2 > 1 - c^2 > 1 - b^2$$

$\therefore \qquad\qquad \dfrac{1}{\sqrt{1 - a^2}} < \dfrac{1}{\sqrt{1 - c^2}} < \dfrac{1}{\sqrt{1 - b^2}}$

i.e. $\qquad\qquad \dfrac{1}{\sqrt{1 - a^2}} < \dfrac{\sin^{-1} b - \sin^{-1} a}{b - a} < \dfrac{1}{\sqrt{1 - a^2}}$

Hence, $\qquad \dfrac{b - a}{\sqrt{1 - a^2}} < \sin^{-1} b - \sin^{-1} a < \dfrac{b - a}{\sqrt{1 - b^2}}$

Example 1.21 : *Prove that :* $\dfrac{b - a}{1 + b^2} < \tan^{-1} b - \tan^{-1} a < \dfrac{b - a}{1 + a^2}$ *if*

$a < b$

Solution : Let $\qquad f(x) = \tan^{-1} x$

$\therefore \qquad\qquad f'(x) = \dfrac{1}{1 + x^2}$ and $f'(c) = \dfrac{1}{1 + c^2}$

By Mean value theorem, we have

$$\frac{\tan^{-1} b - \tan^{-1} a}{b - a} = \frac{1}{1 + c^2}, \quad a < c < b$$

Now, $\qquad\qquad c > a \Rightarrow \dfrac{1}{1 + c^2} < \dfrac{1}{1 + a^2}$

and $$c < b \Rightarrow \frac{1}{1 + c^2} > \frac{1}{1 + b^2}$$

i.e. $$\frac{1}{1 + b^2} < \frac{1}{1 + c^2} < \frac{1}{1 + a^2}$$

i.e. $$\frac{1}{1 + b^2} < \frac{\tan^{-1} b - \tan^{-1} a}{b - a} < \frac{1}{1 + a^2} \quad \text{if } a < b$$

i.e. $$\frac{b - a}{1 + b^2} < \tan^{-1} b - \tan^{-1} a < \frac{b - a}{1 + a^2} \quad \text{if } a < b$$

Example 1.22 : *Use Mean Value Theorem to find approximate value of* $\sqrt{85}$.

Solution : Let $f(x) = \sqrt{x}$ and let $a = 81$ and $b = 85$.

To find the approximate value of $\sqrt{85}$, we apply mean value theorem.

$$f(x) = \sqrt{x} \quad \therefore f'(x) = \frac{1}{2\sqrt{x}} \text{ and } f'(c) = \frac{1}{2\sqrt{c}}$$

$\therefore$
$$\frac{\sqrt{85} - \sqrt{81}}{85 - 81} = \frac{1}{2\sqrt{c}}$$

i.e.
$$\sqrt{85} - \sqrt{81} = \frac{4}{2\sqrt{c}} \quad \text{for some c with } 81 < c < 85$$

We have,
$$81 < c < 85$$

$$\Rightarrow \quad \sqrt{81} < \sqrt{c} < \sqrt{85} < \sqrt{100}$$

$$\Rightarrow \quad 9 < \sqrt{c} < 10$$

$$\Rightarrow \quad \frac{1}{9} > \frac{1}{\sqrt{c}} > \frac{1}{10}$$

i.e.
$$\frac{1}{10} < \frac{1}{\sqrt{c}} < \frac{1}{9}$$

$$\Rightarrow \quad \frac{4}{2\,(10)} < \frac{4}{2\sqrt{c}} < \frac{4}{2\,(9)}$$

i.e.
$$\frac{4}{2\,(10)} < \sqrt{85} - \sqrt{81} < \frac{4}{2\,(9)}$$

$$0.2 < \sqrt{85} - 9 < 0.2222$$

$$9.2 < \sqrt{85} < 9.2222$$

We can improve this result by making use of the result $\sqrt{85} < 9.2222$.

$$\therefore \qquad \frac{4}{2\sqrt{c}} = \frac{4}{2\,(9.2222)} = 0.2168$$

So our improved estimate of $\sqrt{85}$ is

$$9.2168 < \sqrt{85} < 9.2222$$

Think Over It

1. Using mean value theorem, prove that,

   (a) $\dfrac{1}{2\sqrt{n+1}} < \sqrt{n+1} - \sqrt{n} < \dfrac{1}{2\sqrt{n}}, \, n \in \mathbb{N}$

   (b) If f is differentiable on whole $\mathbb{R}$, and f'(x) is constant, then f is a linear function.

   (c) If f is continuous on [a, b] and f' exists and is bounded on the interior, then f is of bounded variation on [a, b].

   (d) If the average speed of a car between two locations was V km/h, then there was at least one instant where the speed indicator displayed V km/h.

2. Suppose $f : [a, b] \times [c, d] \to \mathbb{R}$ is a continuous function with $\dfrac{\partial f}{\partial x}$ continuous and $F(x) = \displaystyle\int_{c}^{d} f(x, y)\, dy$. Prove that F is differentiable and $F'(x) = \displaystyle\int_{c}^{d} \dfrac{\partial f\,(x, y)}{\partial x}\, dy$.

3. If A is an open set in $\mathbb{R}^2$ and $f : A \to \mathbb{R}$ is a function of class C^2, then prove that for each $a \in A$, $\dfrac{\partial^2 f}{\partial x\, \partial y}(a) = \dfrac{\partial^2 f}{\partial y\, \partial x}(a)$.

Summary

1. f is differentiable at c if $\displaystyle\lim_{x \to c} \dfrac{f(x) - f(c)}{x - c} = f'(c)$ exists.

2. If the function f is differentiable at c, then it is continuous at c, but not conversely.

3. If f and g are differentiable at c, then $f \pm g$, $f \cdot g$, αf are differentiable at c.

4. If $g(x) \neq 0$, then $\dfrac{f}{g}$ is differentiable at c.

5. If f is differentiable at c and g is differentiable at $d = f(c)$, then the composite function $g \circ f$ is differentiable at c and

$$(g \circ f)'(c) = g'c(c) \cdot f'(c).$$

6. If f is monotonic function and differentiable at c with $f'(c) \neq 0$, then f^{-1} is differentiable at $d = f(c)$ and $\dfrac{d}{dx} f^{-1}(d) = \dfrac{1}{f'(c)}$.

7. If f is continuous on closed and bounded interval (a, b), differentiable on (a, b), then there exists $c \in (a, b)$, such that

$$f'(c) = \frac{f(b) - f(a)}{b - a}.$$

8. If $f'(x) = 0 \ \forall \ x$ in its domain, then f is a constant function.

9. If $f'(x) \geq 0 \ \forall \ x$, then f is monotonic increasing function.

10. If $f'(x) \leq 0 \ \forall \ x$, then f is monotonic decreasing.

Exercise

[A] Say True or False : Justify !

1. If $f : A \leq \mathbb{R} \to \mathbb{R}$ is continuous, then it is differentiable on A.

2. If $f : I \to \mathbb{R}$ is differentiable, then f is bounded on I, where I is an interval.

3. If $f : [a, b] \to \mathbb{R}$ is differentiable, then f is bounded.

4. Suppose that $A \subseteq \mathbb{R}$ and $f : A \to \mathbb{R}$ is a differentiable function. If $f'(x) = 0$ for all $x \in A$, then f is a constant function.

5. Let $I = [a, b]$ and $f : [a, b] \to \mathbb{R}$ be a function. If f satisfies all the conditions in Rolle's mean value theorem, then there is unique $c \in (a, b)$, such that $f'(c) = 0$.

6. The condition in Rolle's mean value theorem that, f is differentiable on (a, b), is a necessary condition.

7. The condition in the Lagrange's mean value theorem that, f is differentiable on (a, b) is a sufficient condition.

[B] Multiple Choice Questions : Choose the Correct Alternative

1. If $y = \log \tan x$, then $\dfrac{dy}{dx} = $ ......

   (a) cosec x   (b) sec x

   (c) 2 cosec 2x   (d) none of these

2. If $y = \log \cot x$ then $\dfrac{dy}{dx} = $ ......

   (a) sec x   (b) cosec x

   (c) cosec 2x   (d) none of these

3. If $y = \log |x|$, then $\dfrac{dy}{dx} = $ ......

   (a) $-\dfrac{1}{x}$ (b) $\dfrac{1}{|x|}$

   (c) $\dfrac{1}{x}$ (d) none of these

4. If $y = \log (\log x)$ then $\dfrac{dy}{dx}$ at $x = e$ is ......

   (a) 0   (b) e

   (c) e^2 (d) $\dfrac{1}{e}$

5. If $y = x^x$, then $\dfrac{dy}{dx} = $ ......

   (a) $x^x \log x$ (b) $x\, x^{x-1}$

   (c) $x^x (1 + \log x)$ (d) none of these

6. Geometrically Rolle's theorem states that under the given conditions there exists at least one point within the interval (a, b) such that ......

   (a) tangent at that point is parallel to y-axis

   (b) tangent at that point is parallel to x-axis

   (c) tangent at that point is parallel to chord AB

   (d) none of these

7. If f is continuous in [a, b] and differentiable in (a, b), then there exists at least one point c in (a, b) such that $f'(c)$ is equal to ......

   (a) 0   (b) $\dfrac{f(b) + f(a)}{b + a}$

   (c) $\dfrac{f(b) - f(a)}{b - a}$ (d) none of these

8. The value of c of the Rolle's theorem if $f(x) = 2x^3 + x^2 - 4x - 2$ on $[-\sqrt{2}, \sqrt{2}]$ is ......

   (a) 1
   (b) −1
   (c) $\dfrac{3}{2}$
   (d) none of these

9. The value of c of the Rolle's theorem if, $f(x) = \sin x$ in $[0, \pi]$ is ...

   (a) $\dfrac{\pi}{3}$
   (b) $\dfrac{\pi}{4}$
   (c) $\dfrac{\pi}{2}$
   (d) none of these

10. If $f(x) = e^x$ and $g(x) = e^{-x}$ in Cauchy's mean value theorem defined on [a, b] then c is ......

   (a) $\dfrac{a+b}{4}$
   (b) $\dfrac{a+b}{3}$
   (c) $\dfrac{a+b}{2}$
   (d) none of these

[C] Theory Questions :

1. Define differentiability of a function.

2. Prove that every differentiable function is continuous. Is the converse true ?

3. Prove that if $f : I \to \mathbb{R}$ is differentiable with $f'(x) = 0 \ \forall \ x \in I$, then $f(x) = c$ for every $x \in I$.

4. State whether the following statement is true. Justify

   "Let $A \subseteq R$, and $f : A \to \mathbb{R}$. If $f'(x) = 0 \ \forall \ x \in A$, f is a constant function".

5. State and prove the Caratheodary's theorem for differentiability of a function.

6. State and prove the Rolle's mean value theorem.

7. State and prove the Lagrange's mean value theorem.

8. State and prove the Cauchy's mean value theorem.

9. Suppose $f : A \to \mathbb{R}$ is a differentiable function such that $f'(x) \geq 0$. Then prove that f is an increasing function.

10. Show that (i) $|\sin x - \sin y| \leq |x - y|$ for all $x, y \in \mathbb{R}$.

    (ii) $|\cos x - \cos y| \leq |x - y|$ for all $x, y \in \mathbb{R}$.

[D] Numerical Problems :

1. Use the definition to find the derivatives of the following functions :

   (i) $f(x) = x^3$ for $x \in R$

   (ii) $f(x) = \dfrac{1}{x}$ for $x \in R,\ x \neq 0$

   (iii) $f(x) = \dfrac{1}{\sqrt{x}}$ for $x > 0$.

2. If $f : R \to R$ is differentiable at x_0 and that $f(x_0) = 0$, show that $g(x) = |f(x)|$ is differentiable at x_0 iff $f'(x_0) = 0$.

3. If $f(x) = x|x|$, prove that
   $f''(x) = 2$ if $x > 0$ and $f''(x) = -2$ if $x < 0$.

4. If $f(x) = 1 - x$ if $x < 1$
   $$= (1 - x)^2 \quad \text{if } x > 1$$

   Prove that $f'_{+}(1) = 0$ and $f'_{-}(1) = -1$.

5. If $f : R \to R$ is defined by $f(x) = |x|$ show that the function is not derivable at $x = 0$ but is derivable at every other point of its domain.

6. Differentiate the following functions :

   (i) $[(5x + 2)^2 + 3]^4$ (ii) $(x^3 + 2x^2 - 5)^{50}$

   (iii) $[(7x + 4)^4 + (7x + 4)^{-4}]^7$ (iv) $(x^{5/6} + \sqrt{x})^{1/11}$

   (v) $(\sqrt[5]{x} - \sqrt[9]{x})\, x^2$ (vi) $\sin x \cos^{-1} (x^3 + 2)$

   (vii) $\cos^{-1} \sqrt{x}$ (viii) $\dfrac{\cot^{-1} (x + 1)}{\tan^{-1} (x + 1)}$

7. Test whether the conditions of Rolle's theorem hold for the given function on the given interval and if so find the value of 'c'.

   (i) $f(x) = \log\left[\dfrac{(x^2 + ab)}{(a + b)\, x}\right]$ on $[a, b]$.

   (ii) $f(x) = (x - a)^m (x - b)^n$ on $[a, b]$, $m,\ n$ being positive integers.

   (iii) $f(x) = 2 + (x - 1)^{2/3}$ on $[0, 2]$.

   (iv) $f(x) = \dfrac{\sin x}{e^x}$ on $[0, \pi]$

   (v) $f(x) = x(x - 2)\, e^{-x}$ on $[0, 2]$

(vi) $f(x) = \sqrt{(x-a)(b-x)}$ on $[a, b]$

(vii) $f(x) = x(x+3)\, e^{\frac{1}{2}x}$ on $[-3, 0]$.

(viii) $f(x) = |x|$ on $[-1, 1]$.

8.   Test whether the conditions of mean value theorem hold for the given functions on the given interval and if so find the value of 'c'.

(i) $f(x) = (x-1)(x-2)(x-3)$ on $[0, 4]$

(ii) $f(x) = lx^2 + mx + n$ on $[a, b]$

(iii) $f(x) = e^x$ on $[0, 1]$

(iv) $f(x) = \tan^{-1} x$ on $[0, 1]$

(v) $f(x) = x^2 - 3x - 1$ on $[-11/7,\ 13/7]$

9.   If in the Cauchy's mean value theorem

(i) $f(x) = x^2$ and $g(x) = x$

(ii) $f(x) = \sin x$ and $g(x) = \cos x$

Show that in each case 'c' is the arithmetic mean between a and b.

10. If in the Cauchy's mean value theorem.

(i) $f(x) = \sqrt{x}$ and $g(x) = \dfrac{1}{\sqrt{x}}$

Show that 'c' is the geometric mean between a and b.

11. Separate the intervals in which the polynomial
$2x^3 - 15x^2 + 36x + 1$ is increasing or decreasing.

12. For each of the following functions find the points of relative extrema the intervals on which the function is increasing and those on which it is decreasing.

(i) $f(x) = 2x^3 - 9x^2 - 24x + 6$

(ii) $f(x) = 2x^3 + 15x^2 + 24x + 1$

(iii) $f(x) = x^2 - 3x + 5$

(iv) $f(x) = x^3 - 3x - 4$

13. Show that :

(i) $\dfrac{x}{1+x} < \log(1+x) < x, \quad \forall\, x > 0$

(ii) $\dfrac{1}{1+x^2} < \dfrac{\tan^{-1} x}{x} < 1, \quad \forall\, x > 0$

(iii) $x - \dfrac{x^2}{2} + \dfrac{x^3}{3(1+x)} < \log(1+x) < x - \dfrac{x^2}{2} + \dfrac{x^3}{3}, \quad \forall\, x > 0$

14. Show that $|\tan^{-1} x - \tan^{-1} y| \leq |x - y|$

15. Show that $x \leq \sin^{-1} x \leq \dfrac{x}{\sqrt{1 - x^2}}$, $0 \leq x \leq 1$.

16. Use mean value theorem to approximate

    (i) $\sqrt{55}$ (ii) $\sqrt{69}$ (iii) $\sqrt{105}$

17. A function $f(x)$ is continuous in the interval $[0, 1]$ and differentiable in $(0, 1)$. Prove that $f'(c) = f(1) - f(0)$, where $0 < c < 1$.

18. Explain why $1 - |x - 1|$ does not satisfy Rolle's theorem in $[0, 2]$.

19. Use mean value theorem to deduce that $|\sin x - \sin y| < |x - y|$.

20. If $f(x) = (x - 3) \log x$, then show that the equation $x \log x = 3 - x$ is satisfied by at least one value of x lying between 1 and 3.

21. Find the value of 'c' in Cauchy's mean value theorem for the following, if exists

$$f(x) = \log (1 + x), \; g(x) = \log (1 - x) \text{ in } \left[0, \frac{1}{2}\right]$$

22. Show that $f(x) = x^3 - 3x^2 + 3x + 2$ is strictly increasing in every interval.

23. Show that if

$$\frac{a_0}{n + 1} + \frac{a_1}{n} + \dots \frac{a_{n-1}}{2} + a_n = 0, \text{ then the equation}$$

$$a_0 x^n + a_1 x^{n-1} + \dots + a_{n-1} x + a_n = 0, \text{ has at least one root}$$
between 0 and 1.

24. Show that the equation $ax^2 + bx = \dfrac{a}{3} + \dfrac{b}{2}$ has one root between 0 and 1.

25. Show that $-x \leq \sin x \leq x$ for all $x > 0$.

26. Let $a_1, a_2, \dots, a_n$ be real numbers and $f(x) = (a_1 - x)^2 + (a_2 - x)^2 + \dots + (a_n - x)^2$ for $x \in R$ then show that there is a unique point of relative minimum for the function.

27. Verify the Rolle's Theorem for the following functions :

    (a) $f(x) = e^{-x} (\sin x - \cos x)$ in $\left[\dfrac{\pi}{4}, \dfrac{5\pi}{4}\right]$

    (b) $f(x) = x(x - 1)(x - 2)$ in $[0, 2]$

    (c) $f(x) = x(x + 3) e^{-\frac{1}{2}}$ in $[-3, 0]$

Discuss the applicability of Rolle's Mean Value Theorem for

$$f(x) = \begin{cases} x^2 + 1 & \text{for } 0 \le x \le 1 \\ 3 - x & \text{for } 1 \le x \le 2 \end{cases}$$

28. Prove that :

   (a) $\dfrac{x}{1+x} < \log(1+x) < x, \ \forall\, x > 0$

   (b) $\log x < x < \tan x, \ \forall\, x > 1$

   (c) $x - \dfrac{x^2}{2} < \log(1+x) < x - \dfrac{x^2}{2(1+x)}, \ \forall\, x > 0$

   (d) $\dfrac{\tan x}{x} > \dfrac{x}{\sin x}$, for $0 < x < \dfrac{\pi}{2}$.

   (e) $\dfrac{b-a}{1+b^2} < (\tan^{-1} b - \tan^{-1} a) < \dfrac{b-a}{1+a^2}$, where $0 < a < b$ and

   hence deduce that $\dfrac{\pi}{4} + \dfrac{3}{25} < \tan^{-1}\dfrac{4}{3} < \dfrac{\pi}{4} + \dfrac{1}{6}$.

29. (a) Consider the functions $f(x) = \sqrt{x}$, $g(x) = \dfrac{1}{\sqrt{x}}$. Prove that 'c'

   is the geometric mean between a and b by Cauchy's Mean Value Theorem.

   (b) Using function $f(x) = x^{1/x}$, $x > 0$, determine the bigger of two numbers e^x and π^e.

   (c) If $f''(x)$ exists for all points in $[a, b]$ and
   $\dfrac{f(c) - f(a)}{c - a} = \dfrac{f(b) - f(c)}{b - c}$, when $a < c < b$ then $f''(x) = 0$.

   (d) Prove that between any two real roots of $e^x \sin x = 1$, there is at least one root of $e^x(\cos x + \sin x) = 0$.

   (e) Using Lagrange's Mean Value Theorem, prove $x > 0$,
   $$\log_{10}(x+1) = \dfrac{x \log_{10} e}{1 + \theta x} \text{ where, } 0 < \theta < 1.$$

   (f) Prove that, if $a < 1$, $b < 1$ and $a < b$, then
   $$\dfrac{b-a}{\sqrt{1-a^2}} < (\sin^{-1} b - \sin^{-1} a) < \dfrac{b-a}{\sqrt{1-b^2}} \text{ and hence show that}$$
   $$\dfrac{\pi}{6} - \dfrac{1}{2\sqrt{3}} < \sin^{-1}\dfrac{1}{4} < \dfrac{\pi}{6} - \dfrac{1}{\sqrt{15}}.$$

   (g) If $\dfrac{\sin\theta}{\theta} = \cos x\theta$ be equation in x. Show that it has one root

   lying between 0 and 1 and deduce that $\cos\theta < \dfrac{\sin\theta}{\theta} < 1$.

Answers

[A] (1) False   (2) False   (3) True   (4) False   (5) False   (6) False
(7) True

[B] (1) - (c)   (2) - (d)   (3) - (c)   (4) - (d)   (5) - (c)   (6) - (b)   (7) - (c)
(8) - (b)   (9) - (c)   (10) - (c)

[D]

7. (i) $\sqrt{ab}$,   (ii) $\dfrac{na + mb}{m + n}$,   (iii) Not applicable,

(iv) $\dfrac{\pi}{4}$, (v) $2 - \sqrt{2}$, (vi) $\dfrac{a + b}{2}$, (vii) $- 2$.

8. (i) $\dfrac{6 \pm 2\sqrt{3}}{8}$,   (ii) $\dfrac{a + b}{2}$,   (iii) $\log(e - 1)$, (iv) $\sqrt{\dfrac{(4 - \pi)}{\pi}}$,

(v) $\dfrac{1}{7}$

11. Strictly increasing in $(-\infty, 2]$ and $[3, \infty)$ and strictly decreasing in $[2, 3]$.

12. (i)   At $x = -1$, relative maximum

At $x = 4$, relative minimum

Increasing on $(4, \infty)$ and $(-\infty, -1)$

Decreasing on $(-1, 4)$.

(ii)   At $x = -1$, relative minimum

At $x = -4$, relative maximum.

Increasing on $(-1, \infty)$ and $(-\infty, -4)$

Decreasing on $(-4, -1)$.

(iii) Increasing on $(3/2, \infty)$, Decreasing on $(-8, 3/2)$.

(iv) Increasing on $(-\infty, -1)$ and $(1, \infty)$.

16. (i) $7.4038 < \sqrt{55} < 7.4285$

(ii) $8.3007 < \sqrt{69} < 8.3125$

(iii) $10.2439 < \sqrt{105} < 10.25$

Chapter 2...

L'Hospital's Rule and Successive Differentiation

Brook Taylor

Brook Taylor (18 August 1685 - 29 December 1731) was an English mathematician who is best known for **Taylor's theorem** and the **Taylor series.**

Having studied **mathematics** under **John Machin** and **John Keill**, in 1708 he obtained a solution of the problem of the "centre of oscillation," which, remained unpublished until May 1714, when his claim to priority was disputed by **Johann Bernoulli.** Taylor's **Methodus Incrementorium Directa et Inversa (1715)** added a new branch to higher mathematics, now called the **calculus of finite differences".** Among other applications, he used it to determine the form of movement of a vibrating string, by him first successfully reduced to mechanical principles. The same work contained the well-known formula known as **Taylor formula** the importance of which remained unrecognized until 1772, when **J. L. Lagrange** realized its usefulness and termed it "the main foundation of differential calculus".

Taylor's fragile health gave way; he fell into a decline, and died aged 46, on 29 December 1731.

2.1 Indeterminate Forms and L'Hospital's Rules

If $\lim_{x \to a} f(x) = l$ and $\lim_{x \to a} g(x) = m$ and if $m \neq 0$, then we know that,

$$\lim_{x \to a} \frac{f(x)}{g(x)} = \frac{\lim_{x \to a} f(x)}{\lim_{x \to a} g(x)} = \frac{l}{m}$$

But if $l \neq 0$ and $m = 0$, then $\lim\limits_{x \to a} \dfrac{f(x)}{g(x)}$ does not exists.

We now consider the case when $l = 0$ and $m = 0$. In this case, the limit of the quotient $\dfrac{f(x)}{g(x)}$ is said to be indeterminate form of the type $0/0$. Here the limit may be any real value or may not exist.

For example, the fraction $\dfrac{x^2 - 4}{x - 2}$ takes the indeterminate form $0/0$ when $x \to 2$.

But $\lim\limits_{x \to 2} \dfrac{x^2 - 4}{x - 2} = 4$ which shows that the limit exists.

Similarly, if $\lim\limits_{x \to a} f(x) = \infty$ and $\lim\limits_{x \to a} g(x) = \infty$, then in this case, the limit of the quotient $\dfrac{f(x)}{g(x)}$ is said to be indeterminate form of the type ∞/∞.

The other indeterminate forms are represented by the symbols $\infty - \infty$, $0.\infty$, 1^{∞}, 0^{0} and ∞^{0}. By simple adjustment these indeterminate forms can be reduced to either $0/0$ form or ∞/∞ form.

In this section we shall give methods called L'Hospital's rules to find the limits of the functions, which take indeterminate forms.

$\boxed{\textbf{Theorem 1}}$ **(L'Hospital's rule, The case 0/0) :** If $\lim\limits_{x \to a} f(x) = 0$ and $\lim\limits_{x \to a} g(x) = 0$, then, $\lim\limits_{x \to a} \dfrac{f'(x)}{g'(x)} = l \Rightarrow \lim\limits_{x \to a} \dfrac{f(x)}{g(x)} = l$ ■

Remark : If $\lim\limits_{x \to a} \dfrac{f'(x)}{g'(x)}$ also becomes indeterminate form then the above rule can be generalised i.e. if $\lim\limits_{x \to a} f'(x) = 0$ and $\lim\limits_{x \to a} g'(x) = 0$.

Then, $\lim\limits_{x \to a} \dfrac{f''(x)}{g''(x)} = l \Rightarrow \lim\limits_{x \to a} \dfrac{f'(x)}{g'(x)} = l$

This process can be repeated till indeterminate form persists.

$\boxed{\textbf{Theorem 2}}$ **(L'Hospital's rule, The case ∞/∞) :** If $\lim\limits_{x \to a} f(x) = \infty$ and $\lim\limits_{x \to a} g(x) = \infty$, then $\lim\limits_{x \to a} \dfrac{f'(x)}{g'(x)} = l \Rightarrow \lim\limits_{x \to a} \dfrac{f'(x)}{g(x)} = l$ ■

Illustrative Examples

Example 2.1 : $\lim\limits_{x \to 0} \dfrac{x^2 + 2\cos x - 2}{x \sin^3 x}$

Solution : The factor $\sin^3 x$ in the denominator is continuously differentiating. So we write the example as

$$\lim_{x \to 0} \frac{x^2 + 2\cos x - 2}{x^4} \left(\frac{x^3}{\sin^3 x} \right)$$

$$= \lim_{x \to 0} \left(\frac{x^2 + 2\cos x - 2}{x^4} \right) \cdot \lim_{x \to 0} \left(\frac{x^3}{\sin^3 x} \right)$$

$$= \lim_{x \to 0} \left(\frac{x^2 + 2\cos x - 2}{x^4} \right) \cdot 1 \qquad \text{(0/0 form)}$$

$$\therefore \quad \lim_{x \to 0} \left(\frac{x^2 + 2\cos x - 2}{x^4} \right)$$

$$= \lim_{x \to 0} \left(\frac{2x - 2\sin x}{4x^3} \right) \qquad \text{(0/0 form)}$$

$$= \lim_{x \to 0} \left(\frac{2 - 2\cos x}{12x^2} \right) \qquad \text{(0/0 form)}$$

$$= \lim_{x \to 0} \frac{2\sin x}{24x} \qquad \text{(0/0 form)}$$

$$= \lim_{x \to 0} \frac{2\cos x}{24} = \frac{1}{12}$$

Example 2.2 : *Evaluate* $\lim\limits_{x \to 0} \dfrac{\log \sin x}{\cot x}$. **(April 12, 14, Oct. 12, 13)**

Solution : $\lim\limits_{x \to 0} \dfrac{\log \sin x}{\cot x}$ $\qquad (\infty/\infty \text{ form})$

$$= \lim_{x \to 0} \frac{\cos x/\sin x}{-\operatorname{cosec}^2 x} = \lim_{x \to 0} \left(-\frac{\cos x}{\sin x} \sin^2 x \right)$$

$$= \lim_{x \to 0} \left(-\sin x \cos x \right) = 0$$

Example 2.3 : *Evaluate* $\lim\limits_{x \to 0} \left(\dfrac{a}{x} - \cot \dfrac{x}{a} \right)$.

Solution : We write $\dfrac{x}{a} = y$

$\therefore$ As $x \to 0,\ y \to 0$

$$\therefore \quad \lim_{x \to 0} \left(\frac{a}{x} - \cot \frac{x}{a} \right) = \lim_{y \to 0} \left(\frac{1}{y} - \frac{1}{\tan y} \right) \qquad (\infty - \infty \text{ form})$$

$$= \lim_{y \to 0} \frac{\tan y - y}{y \tan y}$$

$$= \lim_{y \to 0} \left(\frac{\tan y - y}{y^2} \right) \left(\frac{y}{\tan y} \right)$$

$$= \lim_{y \to 0} \left(\frac{\tan y - y}{y^2} \right) 1. \qquad (0/0 \text{ form})$$

$$= \lim_{y \to 0} \frac{\sec^2 y - 1}{2y} \qquad (0/0 \text{ form})$$

$$= \lim_{y \to 0} \frac{2 \sec^2 y \tan y}{2} = 0$$

Example 2.4 : *Evaluate* $\lim\limits_{x \to 1} (1 - x) \tan \dfrac{\pi x}{2}$.

Solution : $\lim\limits_{x \to 1} (1 - x) \tan \dfrac{\pi x}{2}$ $(0.\infty \text{ form})$

$$= \lim_{x \to 1} \frac{(1 - x)}{\cot \dfrac{\pi x}{2}} \qquad (0/0 \text{ form})$$

$$= \lim_{x \to 1} \frac{-1}{-\dfrac{\pi}{2} \operatorname{cosec}^2 \dfrac{\pi x}{2}} = \lim_{x \to 1} \frac{2}{\pi} \sin^2 \frac{\pi x}{2} = \frac{2}{\pi}$$

Example 2.5 : *Evaluate* $\lim\limits_{x \to 1} x^{\left(\frac{1}{1-x} \right)}$

Solution : $\lim\limits_{x \to 1} x^{\left(\frac{1}{1-x} \right)}$

$$\text{Let } y = \lim_{x \to 1} x^{\left(\frac{1}{1-x}\right)}$$

$$\therefore \quad \log y = \lim_{x \to 1} \frac{1}{(1-x)} \log x$$

$$\left(\lim_{x \to a} \log f(x) = \log \lim_{x \to a} f(x) \right)$$

$$= \lim_{x \to 1} \frac{\log x}{(1-x)} \qquad (0/0 \text{ form})$$

$$= \lim_{x \to 1} \frac{1/x}{-1} = -1$$

$$\therefore \quad \log y = -1$$

$$\therefore \quad y = e^{-1} = \frac{1}{e}$$

Example 2.6 : *Evaluate* $\lim_{x \to 0} x^x$

Solution : $\lim_{x \to 0} x^x$ (0^0 form)

Let $y = \lim_{x \to 0} x^x$

$$\therefore \quad \log y = \lim_{x \to 0} x \log x \qquad (0.\infty \text{ form})$$

$$= \lim_{x \to 0} \frac{\log x}{1/x} \qquad (\infty/\infty \text{ form})$$

$$= \lim_{x \to 0} \left(\frac{\dfrac{1}{x}}{-\dfrac{1}{x^2}} \right) = \lim_{x \to 0} (-x) = 0$$

$$\therefore \quad \log y = 0$$

$$\therefore \quad y = e^0 = 1$$

Example 2.7 : *Evaluate* $\lim_{x \to \frac{\pi}{2}} (sec\ x)^{cot\ x}$.

Solution : $\lim_{x \to \frac{\pi}{2}} (\sec x)^{\cot x}$ $(\infty^0 \text{ form})$

Let
$$y = \lim_{x \to \frac{\pi}{2}} (\sec x)^{\cot x}$$

$\therefore$
$$\log y = \lim_{x \to \frac{\pi}{2}} \cot x \cdot \log \sec x \qquad (0.\infty \text{ form})$$

$$= \lim_{x \to \frac{\pi}{2}} \frac{\log \sec x}{\tan x} \qquad (\infty/\infty \text{ form})$$

$$= \lim_{x \to \frac{\pi}{2}} \frac{\dfrac{1}{\sec x} \cdot \sec x \cdot \tan x}{\sec^2 x}$$

$$= \lim_{x \to \frac{\pi}{2}} \left(\frac{\tan x}{\sec^2 x} \right) = \lim_{x \to \frac{\pi}{2}} (\sin x \cdot \cos x) = 0$$

$$\log y = 0$$

$\therefore$
$$y = e^0 = 1$$

2.2 Successive Differentiation

If $y = f(x)$, and if it is differentiable on an interval then the first derivative of $f(x)$ is given by $f'(x)$. If $f'(x)$ is also differentiable then the derivative of $f'(x)$ is given by $f''(x)$. In a similar manner the nth derivative of $f(x)$ is given by $f^n(x)$.

The successive derivatives of $f(x)$ may be represented by the following symbols :

$f'(x),$	$f''(x),$	$f'''(x)$	......	$f^n(x)$ ...
or $y',$	$y'',$	y'''	......	$y^{(n)}$ ...
or $y_1,$	$y_2,$	y_3	......	y_n ...
or $\dfrac{dy}{dx},$	$\dfrac{d^2y}{dx^2},$	$\dfrac{d^3y}{dx^3}$	......	$\dfrac{d^ny}{dx^n}$ ...
or $Dy,$	$D^2y,$	D^3y	......	D^ny ...

The symbol $f^n(c)$ or $\left[\dfrac{d^ny}{dx^n} \right]_{x=c}$ or $y_n(c)$ denotes the value of nth derivatives at $x = c$.

2.2.1 n^{th} Derivatives of Some Standard Functions

1. If $\qquad y = (ax + b)^m$

   then, $\qquad y_1 = ma(ax + b)^{m-1}$

   $\qquad\qquad y_2 = m(m - 1)\, a^2\, (ax + b)^{m-2}$

   $\qquad\qquad y_3 = m(m - 1)\, (m - 2)\, a^3\, (ax + b)^{m-3}$

By induction, we can prove that,

$$y_n = m(m - 1)\, (m - 2) \ldots (m - n + 1)$$
$$a^n\, (ax + b)^{m-n},\ \forall\, n \geq 1.$$

If m is a positive integer, then y_n can be written as

$$y_n = \frac{m!}{(m - n)!}\, a^n\, (ax + b)^{m-n} \qquad \text{if } n < m$$

$$= m!\, a^m \qquad\qquad\qquad \text{if } n = m \qquad \ldots \text{(i)}$$

$$= 0 \qquad\qquad\qquad\qquad \text{if } n > m$$

In particular, if $\qquad y = x^m$

then, $\qquad\qquad y_n = \dfrac{m!}{(m - n)!}\, x^{m-n} \qquad\qquad \text{if } n < m$

$$= m! \qquad\qquad\qquad \text{if } n = m$$

$$= 0 \qquad\qquad\qquad\quad \text{if } n > m$$

2. If $\qquad y = \dfrac{1}{(ax + b)}$

   put $\qquad\quad m = -1 \text{ in (i)}$

   $\therefore \qquad\quad y_n = (-1)\, (-2)\ \ldots\ (-n)\ a^n\, (ax + b)^{-1-n}$

   $$= \frac{(-1)^n\, (n!)\, a^n}{(ax + b)^{n+1}}, n \in \mathbb{N} \qquad\qquad \ldots \text{(ii)}$$

In particular, if $\qquad y = \dfrac{1}{x}$

   then $\qquad\qquad y_n = \dfrac{(-1)^n\, (n!)}{x^{n+1}}$

3. If $\qquad y = \log (ax + b)$

   then, $\qquad\quad y_1 = \dfrac{a}{(ax + b)}$

$\therefore$ By above result (ii), we have,

$$y_n = \frac{(-1)^{n-1}(n-1)!\, a^n}{(ax+b)^n} \qquad \ldots \text{(iii)}$$

In particular, if $\quad y = \log x$

then $\quad\quad y_n = \dfrac{(-1)^{n-1}(n-1)!}{x^n}$

4. If $\quad\quad\quad y = a^{mx}$

$\therefore \quad\quad\quad y_1 = ma^{mx} . \log a$

$\quad\quad\quad\quad y_2 = m^2\, a^{mx} . (\log a)^2$

$\quad\quad\quad\quad y_3 = m^3\, a^{mx} . (\log a)^3$

By induction, we can prove that,

$$y_n = m^n\, a^{mx}\, (\log a)^n \ \forall\, n \geq 1 \qquad \ldots \text{(iv)}$$

5. If $\quad\quad\quad y = e^{mx}$

By putting $a = e$ in result (iv), we get

$$y_n = m^n\, e^{mx}, \text{ for each } n \qquad \ldots \text{(v)}$$

6. If $\quad\quad\quad y = \sin(ax+b)$

then, $\quad\quad y_1 = a\cos(ax+b) = a\sin\left(ax+b+\dfrac{\pi}{2}\right)$

$\quad\quad\quad\quad y_2 = -a^2\sin(ax+b) = a^2\sin\left(ax+b+\dfrac{2\pi}{2}\right)$

$\quad\quad\quad\quad y_3 = -a^3\cos(ax+b) = a^3\sin\left(ax+b+\dfrac{3\pi}{2}\right)$

By induction, we can prove that,

$$y_n = a^n\sin\left(ax+b+\dfrac{n\pi}{2}\right) \text{ for each } n \geq 1. \ \ldots \text{(vi)}$$

7. If $\quad\quad\quad y = \cos(ax+b)$

then, $\quad\quad y_1 = -a\sin(ax+b) = a\cos\left(ax+b+\dfrac{\pi}{2}\right)$

$\quad\quad\quad\quad y_2 = -a^2\cos(ax+b) = a^2\cos\left(ax+b+\dfrac{2\pi}{2}\right)$

$\quad\quad\quad\quad y_3 = -a^3\sin(ax+b) = a^3\cos\left(ax+b+\dfrac{3\pi}{2}\right)$

$\therefore$ Using induction, we can prove that,

$$y_n = a^n\cos\left(ax+b+\dfrac{n\pi}{2}\right), \text{ for each } n \in N \qquad \ldots \text{(vii)}$$

8.   If $y = e^{ax} \sin (bx + c)$

     then, $y_1 = ae^{ax} \sin (bx + c) + e^{ax} b \cos (bx + c)$

     Let, $a = r \cos \theta$ and $b = r \sin \theta$

     $\therefore$ $r = (a^2 + b^2)^{1/2}$ and $\theta = \tan^{-1} \dfrac{b}{a}$

     $\therefore$ $y_1 = r \cos \theta \, e^{ax} \sin (bx + c) + e^{ax} r \sin \theta \cos (bx + c)$

     $\qquad\qquad = r \, e^{ax} [\cos \theta \sin (bx + c) + \sin \theta \cos (bx + c)]$

     $\qquad\qquad = r \, e^{ax} \sin (bx + c + \theta)$

     and $y_2 = r^2 \, e^{ax} \sin (bx + c + 2\theta)$

     By induction, we can prove that,

     $\qquad\qquad y_n = r^n \, e^{ax} \sin (bx + c + n\theta)$ ... (viii)

9.   If $y = e^{ax} \cos (bx + c)$

     then, by induction, we can prove that

     $\qquad\qquad y_n = r^n \, e^{ax} \cos (bx + c + n\theta)$ for $n \geq 1$ ... (ix)

2.2.2 Leibnitz's Theorem

This theorem helps us to find the nth derivatives of the product of two functions.

Statement : If $y = uv$ where u and v are functions of x possessing derivatives of nth order, then

$$y_n = (uv)_n = n_{c_0} u_n v + n_{c_1} u_{n-1} v_1 + n_{c_2} u_{n-2} v_2 + \ldots n_{c_r} u_{n-r} v_r$$

$$+ \ldots + n_{c_n} u \, v_n \qquad\qquad \text{... (x)}$$

     where, $n_{c_r} = \dfrac{n!}{(n-r)! \, r!}$

Proof : We shall prove this theorem by mathematical induction.

     If $y = uv$

then by direct differentiation, we have,

$$y_1 = u_1 v + u v_1$$

$$= 1_{c_0} u_1 v + 1_{c_1} u v_1 \qquad\qquad \text{... (i)}$$

Hence the result (x) is true for $n = 1$.

Now we assume that the result (x) is true for $n = m$.

$\therefore$ $y_m = (uv)_m = m_{c_0} u_m v + m_{c_1} u_{m-1} v_1 + m_{c_2} u_{m-2} v_2$

$\qquad + \ldots m_{c_{r-1}} u_{m-r+1} v_{r-1} + \ldots + m_{c_r} u_{m-r} v_r + \ldots + m_{c_{m-1}} u_1 v_{m-1}$

$$+ m_{c_m} u v_m$$

Differentiating this result, we get

$$y_{m+1} = (uv)_{m+1} = {}^m c_0 \, u_{m+1} v + {}^m c_0 \, u_m v_1 + {}^m c_1 \, u_m v_1 + {}^m c_1 \, u_{m-1}$$

$$v_2 + {}^m c_2 \, u_{m-1} v_2 + {}^m c_2 \, u_{m-2} v_3 + \ldots + {}^m c_{r-1} \, u_{m-r+2} v_{r-1} + {}^m c_{r-1}$$

$$u_{m-r+1} v_r + {}^m c_r \, u_{m-r+1} v_r + {}^m c_r \, u_{m-r} v_{r+1} + \ldots + {}^m c_m \, u_1 v_m + {}^m c_m$$

$$u \, v_{m+1}$$

$$y_{m+1} = (uv)_{m+1} = {}^m c_0 u_{m+1} v + \left({}^m c_0 + {}^m c_1\right) u_m v_1 +$$

$$\left({}^m c_1 + {}^m c_2\right) u_{m-2} v_2 + \ldots + \left({}^m c_{r-1} + {}^m c_r\right) u_{m-r+1} v_r + \ldots\ldots + {}^m c_m$$

$$u \, v_{m+1}$$

But we have the result

$${}^m c_{r-1} + {}^m c_r = {}^{m+1} c_r \, , \; {}^m c_m = {}^{m+1} c_{m+1} \; \text{and} \; {}^m c_0 = {}^{m+1} c_0$$

$$\therefore \; y_{m+1} = (uv)_{m+1} = {}^{m+1} c_0 \, u_{m+1} v + {}^{m+1} c_1 \, u_m v_1 + {}^{m+1} c_2$$

$$u_{m-2} v_2 + \ldots + {}^{m+1} c_r \, u_{m-r+1} v_r + \ldots + {}^{m+1} c_{m+1} \, uv_{m+1} \quad \ldots \text{(ii)}$$

This result shows that, if the theorem is true for n = m, then it is also true for n = m + 1. Hence, from results (i) and (ii), the theorem is true for all positive integral values of n. Hence, the theorem.

Illustrative Examples

Example 2.8 : *If* $\quad y = \dfrac{x^2 + 4x + 1}{x^3 + 2x^2 - x - 2}$, *find* y_n.

Solution : We have

$$\frac{x^2 + 4x + 1}{x^3 + 2x^2 - x - 2} = \left\{ \frac{1}{x-1} + \frac{1}{x+1} - \frac{1}{x+2} \right\}$$

Applying the result (ii), we get

$$y_n = \frac{(-1)^n \, n!}{(x-1)^{n+1}} + \frac{(-1)^n \, n!}{(x+1)^{n+1}} - \frac{(-1)^n \, n!}{(x+2)^{n+1}}$$

$$= (-1)^n \, n! \cdot \left[\frac{1}{(x-1)^{n+1}} + \frac{1}{(x+1)^{n+1}} - \frac{1}{(x+2)^{n+1}} \right]$$

Example 2.9 : *If* $\quad y = e^{ax} \cos^2 x \sin x$, *find* n^{th} *derivative of* y.

Solution : We have, $y = e^{ax} \cos^2 x \sin x$

$$= e^{ax} \frac{1}{2} (1 + \cos 2x) \, \sin x$$

$$= e^{ax} \left(\frac{1}{2} \sin x + \frac{1}{2} \cos 2x \sin x \right)$$

$$= e^{ax}\left[\frac{1}{2}\sin x + \frac{1}{4}\sin 3x - \frac{1}{4}\sin x\right]$$

$$= e^{ax}\left[\frac{1}{4}\sin x + \frac{1}{4}\sin 3x\right]$$

$$\therefore \qquad y = \frac{1}{4}e^{ax}\sin x + \frac{1}{4}e^{ax}\sin 3x$$

$$\therefore \qquad y_n = \frac{1}{4}(a^2+1)^{n/2}\,e^{ax}\sin\left(x + n\tan^{-1}\frac{1}{a}\right)$$

$$+ \frac{1}{4}(a^2+9)^{n/2}\,e^{ax}\sin\left(3x + n\tan^{-1}\frac{3}{a}\right)$$

Example 2.10 : *If $y = x^2\,e^x\sin x$, find y_n.*

Solution : We take $e^x \sin x$ as the first factor and x^2 as the second factor of y.

$$\therefore \qquad y = e^x \sin x . x^2$$

To find y_n, we use Leibnitz theorem.

$$\therefore \qquad y_n = (e^x \sin x . x^2)_n$$

$$= n_{c_0}(e^x \sin x)_n\, x^2 + n_{c_1}(e^x \sin x)_{n-1}(x^2)_1$$

$$+ n_{c_2}(e^x \sin x)_{n-2}(x^2)_2$$

$$= 2^{n/2}\,e^x \sin(x + n\tan^{-1} 1) . x^2 + n.2^{(n-1)/2}\,e^x \sin(x + (n-1)$$

$$\tan^{-1} 1) . 2x + \frac{n(n-1)}{2}\,2^{(n-2)/2}$$

$$e^x \sin(x + (n-2)\tan^{-1} 1) .2$$

$$= 2^{n/2}\,e^x \sin\left(x + n\frac{\pi}{4}\right)x^2 + 2^{\frac{n}{2}-\frac{1}{2}}\,ne^x \sin\left(x + (n-1)\frac{\pi}{4}\right) . x$$

$$+ 2^{\frac{n}{2}-2}\,n(n-1)\,e^x \sin\left(x + (n-2)\frac{\pi}{4}\right)$$

Example 2.11 : *If $y = (sin^{-1} x)^2$, prove that*

$$(1 - x^2)\,y_{n+2} - (2n+1)\,xy_{n+1} - n^2\,y_n = 0$$

Solution : We have, $y = (\sin^{-1} x)^2$

$$\therefore \qquad y_1 = 2\sin^{-1} x . \frac{1}{\sqrt{1-x^2}}$$

Squaring and simplifying, we have

$$(1-x^2)\,y_1^2 - 4y = 0$$

Differentiating once again we get

$(1 - x^2)\, 2y_1\, y_2 - 2xy_1^2 - 4\, y_1 = 0$

i.e. $2y_1 \left[(1 - x^2)\, y_2 - xy_1 - 2 \right] = 0$

$\Rightarrow \qquad (1 - x^2)\, y_2 - xy_1 - 2 = 0 \qquad\qquad$ ... (i) since $2y_1 \neq 0$

Differentiating (i) n times using Leibnitz's theorem, we have

$y_{n+2}\, (1 - x^2) + ny_{n+1}\, (-2x) + \dfrac{n(n-1)}{2}\, y_n\, (-2) - (y_{n+1}\, x + n\, y_n \cdot 1) = 0$

$(1 - x^2)\, y_{n+2} - 2n\, x\, y_{n+1} - n\,(n-1)\, y_n - x\, y_{n+1} - ny_n = 0$

i.e. $(1 - x^2)\, y_{n+2} - (2n+1)\, x\, y_{n+1} - n^2\, y_n = 0$

Example 2.12 : *If* $\quad y = cos\,(m\,sin^{-1} x)$

show that $(1 - x^2)\, y_{n+2} - (2n+1)\, xy_{n+1} + (m^2 - n^2)\, y_n = 0$

Solution : We have, $y = \cos\,(m\,\sin^{-1} x)$

$$y_1 = -\sin\,(m\,\sin^{-1} x) \cdot m\,\frac{1}{\sqrt{1 - x^2}}$$

$\therefore \qquad \sqrt{1 - x^2}\ y_1 = -m\,\sin\,(m\,\sin^{-1} x)$

Differentiating once again, we get,

$$\sqrt{1 - x^2}\, y_2 + \frac{-2x}{2\sqrt{1 - x^2}}\, y_1 = -m\,\cos\,(m\,\sin^{-1} x)\, \cdot \frac{m}{\sqrt{1 - x^2}}$$

$\therefore \qquad (1 - x^2)\, y_2 - xy_1 + m^2 y = 0 \qquad\qquad$ ... (i)

Differentiating (i) n times using Leibnitz's theorem, we get,

$$y_{n+2}\, (1 - x^2) + ny_{n+1}\, (-2x) + \frac{n(n-1)}{2}\, y_n(-2)$$

$$- (y_{n+1}\, x + n\, y_n \cdot 1) + m^2\, y_n = 0$$

$(1 - x^2)\, y_{n+2} - 2nx\, y_{n+1} - n(n-1)\, y_n - xy_{n+1} - ny_n + m^2\, y_n = 0$

i.e. $(1 - x^2)\, y_{n+2} - (2n+1)\, xy_{n+1} + (m^2 - n^2)\, y_n = 0.$

Example 2.13 : *If* $x = tan\,(log\,y)$*, prove that*

$\qquad (1 + x^2)\, y_{n+1} + (2nx - 1)\, y_n + n\,(n-1)\, y_{n-1} = 0$

Solution : We have, $x = \tan\,(\log y)$

$\therefore \qquad\qquad \log y = \tan^{-1} x$

hence, $\qquad\qquad y = e^{\tan^{-1} x} \qquad\qquad$ ... (i)

Differentiating (i) with respect to x, we get

$$y_1 = e^{\tan^{-1} x}\, \frac{1}{1 + x^2}$$

i.e.　　$(1 + x^2)\, y_1 = e^{\tan^{-1} x} = y$

i.e.　$(1 + x^2)\, y_1 - y = 0$ 　　　　　　　… (ii)

Differentiating (ii) n times using Leibnitz's theorem, we get

$$y_{n+1}\,(1 + x^2) + n y_n\,(2x) + \frac{n(n-1)}{2}\, y_{n-1}\,(2) - y_n = 0$$

i. e. $(1 + x^2)\, y_{n+1} + 2nx y_n + n(n-1)\, y_{n-1} - y_n = 0$

i. e. $(1 + x^2)\, y_{n+1} + (2nx - 1)\, y_n + n\,(n-1)\, y_{n-1} = 0$

Example 2.14 : *If* $y^{1/m} + y^{-1/m} = 2x,$ *show that*

$$(x^2 - 1)\, y_{n+2} + (2n + 1)\, x y_{n+1} + (n^2 - m^2)\, y_n = 0.$$

Solution : We have, 　　$y^{\frac{1}{m}} + \dfrac{1}{y^{\frac{1}{m}}} = 2x$

Put, 　　　　　　　　$y^{\frac{1}{m}} = t$

$\therefore$ 　　　　　　　$t + \dfrac{1}{t} = 2x$

i.e. 　　　　　　$t^2 - 2xt + 1 = 0$

which is quadratic in 't'. Solving it for t, we get

$$t = x \pm \sqrt{x^2 - 1}$$

We ignore the negative sign

$\therefore$ 　　　　　　$t = y^{1/m} = x + \sqrt{x^2 - 1}$

$\therefore$ 　　　　　　$y = \left(x + \sqrt{x^2 - 1}\right)^m$ 　　　… (i)

Differentiating (i) with respect to x, we get

$$y_1 = m\left(x + \sqrt{x^2 - 1}\right)^{m-1}\left[1 + \frac{x}{\sqrt{x^2 - 1}}\right]$$

$\therefore$ 　　　　$y_1 = \dfrac{m\left(x + \sqrt{x^2 - 1}\right)^m}{\sqrt{x^2 - 1}} = \dfrac{my}{\sqrt{x^2 - 1}}$

i.e. 　　$\sqrt{x^2 - 1}\ y_1 = my$

Differentiating once again, we get,

$$\sqrt{x^2 - 1}\ y_2 + y_1 \frac{x}{\sqrt{x^2 - 1}} = my_1 = m\,\frac{my}{\sqrt{x^2 - 1}}$$

$\therefore$ 　　$(x^2 - 1)\, y_2 + x y_1 - m^2 y = 0$ 　　　　… (ii)

Differentiating (ii), n times using Leibnitz's theorem, we get,

$$y_{n+2}\,(x^2-1) + ny_{n+1}\,2x + \frac{n(n-1)}{2}\;y_n\,.\,2 + y_{n+1}\;x + ny_n\,.$$

$$1 - m^2\,y_n = 0$$

i.e. $(x^2-1)\,y_{n+2} + 2nxy_{n+1} + n(n-1)\,y_n + xy_{n+1} + ny_n - m^2\,y_n = 0$

i. e. $(x^2-1)\,y_{n+2} + (2n+1)\;xy_{n+1} + (n^2-m^2)\,y_n = 0$

Example 2.15 : *If* $\cos^{-1}\left(\dfrac{y}{b}\right) = \log\left(\dfrac{x}{n}\right)^n$, *show that,*

$$x^2\,y_{n+2} + (2n+1)\,xy_{n+1} + 2n^2 y_n = 0$$

Solution : We have,

$$\cos^{-1}\left(\frac{y}{b}\right) = \log\left(\frac{x}{n}\right)^n$$

$$\therefore \qquad \cos^{-1}\left(\frac{y}{b}\right) = n\log\left(\frac{x}{n}\right)$$

i.e. $\qquad \cos^{-1}\left(\dfrac{y}{b}\right) = n\,(\log x - \log n)$ ... (i)

Differentiating (i) with respect to x, we get

$$\frac{-1}{\sqrt{1 - \dfrac{y^2}{b^2}}}\cdot\frac{y_1}{b} = \frac{n}{x}$$

i.e. $\qquad y_1^2\,x^2 = n^2\,(b^2 - y^2)$ ... (ii)

Again differentiating (ii) with respect to x,

$$2y_1\,y_2\,x^2 + 2xy_1^2 = -2n^2\,yy_1$$

i.e. $\qquad 2y_1\,(y_2\,x^2 + xy_1 + n^2 y) = 0$

i.e. $\qquad y_2\,x^2 + xy_1 + n^2\,y = 0 \quad (\because 2y_1 \neq 0)$ ... (iii)

Differentiating (iii) n times using Leibnitz's theorem

$$y_{n+2}\,x^2 + ny_{n+1}\,2x + \frac{n(n-1)}{2}\;y_n\,.\,2 + y_{n+1}\;x + ny_n\,.\,1 + n^2\,y_n = 0$$

i.e. $x^2\,y_{n+2} + 2nx\,y_{n+1} + n(n-1)\,y_n + xy_{n+1} + ny_n + n^2\,y_n = 0$

i.e. $x^2\,y_{n+2} + (2n+1)\;xy_{n+1} + 2n^2\,y_n = 0$

Example 2.16 : *If* $y = e^{ax}u$, *where* u *is a function of* x, *then show that*

$$D^n y = e^{ax}\,(D+a)^n\,u$$

Solution : We have, $y = ue^{ax}$ ... (i)

Differentiating (i) n times using Leibnitz's theorem, we get

$$y_n = D^n y = n_{C_0} u_n (e^{ax}) + n_{C_1} u_{n-1} (e^{ax})_1 +$$

$$n_{C_2} u_{n-2} (e^{ax})_2 + \ldots\ldots + n_{C_r} u_{n-r} (e^{ax})_r + \ldots + n_{C_n} u(e^{ax})_n$$

i.e.

$$D^n y = n_{C_0} u_n e^{ax} + n_{C_1} u_{n-1} a\ e^{ax} + \ldots\ldots +$$

$$n_{C_r} u_{n-r} a^r e^{ax} + \ldots\ldots + n_{C_n} u\ a^n e^{ax}$$

But

$$Du = u_1,\ D^2u = u_2,\ \ldots\ D^n u = u_n$$

So,

$$D^n y = n_{C_0} D^n u\ e^{ax} + n_{C_1} D^{n-1} u\ ae^{ax}$$

$$+ n_{C_2} D^{n-2} u\ a^2 e^{ax} + \ldots\ldots + n_{C_r} D^{n-r} u\ a^r e^{ax} + \ldots\ldots$$

$$+ n_{C_n} u\ a^n e^{ax}$$

$$= e^{ax} \left(n_{C_0} D^n u + n_{C_1} D^{n-1} u\ a + n_{C_2} D^{n-2} u\ a^2 \right.$$

$$\left. + \ldots\ldots + n_{C_r} D^{n-r} u\ a^r + \ldots\ldots + n_{C_n} u\ a^n \right)$$

$$= e^{ax} \left(n_{C_0} D^n + n_{C_1} D^{n-1} a + n_{C_2} D^{n-2} a^2 \right.$$

$$\left. + \ldots\ldots + n_{C_r} D^{n-r} a^r + \ldots\ldots + n_{C_n} a^n \right) u$$

Treating D as operator, we have

$$D^n y = e^{ax} (D + a)^n u$$

2.3 Taylor's and Maclaurin's Theorem

Taylor's theorem may be regarded as an extension of mean value theorem to higher order derivatives. Taylor's theorem has many applications in numerical estimation, inequalities, extreme values of a function and convex functions.

Mean value theorem gives the relation between the values of a function and its first derivative, whereas Taylor's theorem provides a relation between the values of a function and its higher derivatives.

Taylor's Theorem with Lagrange's form of Remainder

(Oct. 2012, April 2014)

If a function f defined in [a, b] is such that

(i) f and its derivatives f', f'' … f^{n-1} are continuous in [a, b]

(ii) f^n exists in (a, b) then $\exists\, c \in$ (a, b) such that

$$f(b) = f(a) + (b-a)\,f'(a) + \frac{(b-a)^2}{2!}\,f''(a) + \ldots + \frac{(b-a)^{n-1}}{(n-1)!}\,f^{n-1}(a)$$

$$+ \frac{(b-a)^n}{n!}\,f^n(c) \quad \ldots (1)$$

where, $\dfrac{(b-a)^n}{n!}\,f^n(c) = R_n$ is called Lagrange's form of remainder

after n terms

Another form :

If a function f defined in [a, a + h] is such that

(i) f and its derivatives f', f'', … f^{n-1} are continuous in [a + h]

(ii) f^n exists in (a, a + h) then $\exists$ a real number θ, $0 < \theta < 1$ such

$$f(a + h) = f(a) + hf'(a) + \frac{h^2}{2!}\,f''(a) + \ldots + \frac{h^n}{n!}\,f^n(a + \theta h) \quad \ldots (2)$$

where, $\dfrac{h^n}{n!}\,f^n(a + \theta h) = R_n$ is called Lagrange's form of remainder

after n terms.

Maclaurin's theorem with Lagrange's form of remainder

If a function f defined in [0, x] is such that

(i) f and its derivatives f', f'' … f^{n-1} are continuous in [0, x]

(ii) f^n exists in (0, x)

then $\exists$ a real number θ, $0 < \theta < 1$ such that

$$f(x) = f(0) + f(0) + x\,f'(0) + \frac{x^2}{2!}\,f''(0) + \ldots + \frac{x^{n-1}}{(n-1)!}\,f^{n-1}(0) + \frac{x^n}{n!}\,f^n(\theta x)$$

$$\ldots (3)$$

where $\dfrac{x^n}{n!}\,f^n(x) = R_n$ is called Lagrange's form of remainder after n

terms.

Note : In Taylor's theorem (2), if we put a = 0, h = x, we get Maclaurin's theorem with Lagrange's form of remainder after n terms.

Taylor's and Maclaurin's Series :

In Taylor's theorem (2),

(i) if a function f possesses a derivative of every order in (a, a + h).

(ii) The Lagrange's remainder

$$R_n \;=\; \frac{n^n}{n!}\,f^n(a + \theta h) \longrightarrow 0 \text{ as } n \to \infty$$

then
$$f(a + h) = f(a) + hf'(a) + \frac{h^2}{2!} f''(a) + \ldots + \frac{n^n}{n!} f^n(a) + \ldots$$
$$\ldots (4)$$

The series on R.H.S. of equation (4) is known as Taylor's series

If we put $a = 0$, $h = x$ in equation (4), we get

$$f(x) = f(0) + xf'(0) + \frac{x^2}{2!} f''(0) + \ldots + \frac{x^n}{n!} f(0) + \ldots$$
$$\ldots (5)$$

The series on R.H.S. of (5) is known as **Maclaurin's Series.**

Illustrative Examples

Example 2.17 : *Expand the following with the help of Maclaurin's theorem :*

(i) e^x (ii) sin x (iii) cos x (iv) $(1 + x)^m$ (v) log $(1 + x)$.

Solution : (i) Let
$$f(x) = e^x$$

$\therefore$
$$f'(x) = e^x \qquad\qquad f'(0) = 1$$
$$f''(x) = e^x \qquad\qquad f''(0) = 0$$
$$\vdots \qquad\qquad\qquad \vdots$$
$$f^n(x) = e^x \qquad\qquad f^n(0) = 1$$

Now by Maclaurin theorem with Lagrange's form of remainder

$$f(x) = f(0) + x\, f'(0) + \frac{x^2}{2!} f''(0) + \ldots +$$

$$\frac{x^{n-1}}{(n-1)!} f^{n-1}(0) + \frac{x^n}{n!} f^n(\theta x)$$

where $R_n = \dfrac{x^n}{n!} f^n(\theta x)$ is the remainder after n terms. $R_n \to 0$ as $n \to \infty$

because

$$\lim_{n \to \infty} \left| \frac{R_{n+1}}{R_n} \right| = 0 < 1 \qquad \forall\, x \in \mathbb{R}$$

$$e^x = 1 + x + \frac{x^2}{2!} + \frac{x^3}{3!} + \ldots + \frac{x^{n-1}}{(n-1)!} + \ldots \forall\, x \in \mathbb{R}$$

(ii) Let,
$$f(x) = \sin x$$

$$f'(x) = \cos x = \sin\left(x + \frac{\pi}{2}\right)$$

$$f''(x) = -\sin x = \sin\left(\frac{2\pi}{2} + x\right)$$

$$\vdots$$

$$f^n(x) = \sin\left(\frac{n\pi}{2} + x\right) \forall\, x \in R$$

Also, $\quad f^n(0) = \sin\left(\frac{n\pi}{2}\right) = 0$ if n is even $= \pm 1$, if n is odd.

$\therefore$ By Maclaurin's theorem

$$f(x) = f(0) + x\,f'(0) + \frac{x^2}{2!} f''(0) + \ldots + \frac{x^{n-1}}{(n-1)!}$$

$$f^{n-1}(0) + \frac{x^n}{n!}\frac{x^n}{n!} f^n(\theta x)$$

where, $\dfrac{x^n}{n!} f^n(\theta x) = R_n$ is the remainder after n terms.

$R_n \to 0$ as $n \to \infty$, because

$$\lim_{n\to\infty} \left|\frac{R_{n+1}}{R_n}\right| = 0 < 1 \qquad\qquad \forall\, x \in \mathbb{R}$$

$\therefore \qquad\qquad f(x) = x - \dfrac{x^3}{3!} + \dfrac{x^5}{5!} - \dfrac{x^7}{7!} + \ldots \qquad\qquad \forall\, x \in \mathbb{R}$

(iii) Let, $\qquad\qquad f(x) = \cos x$

$$f'(x) = -\sin x = \cos\left(\frac{\pi}{2} + x\right)$$

$$f''(x) = -\cos x = \cos\left(\frac{2\pi}{2} + x\right)$$

$$\vdots$$

$$f^n(x) = \cos\left(\frac{n\pi}{2} + x\right) \qquad\qquad \forall\, x \in R$$

Also, $\qquad\qquad f^n(0) = \cos\left(\frac{n\pi}{2}\right) = 0$ if n is odd $= \pm 1$ if n is even.

$\therefore$ By Maclaurin theorem

$$f(x) = f(0) + x\,f'(0) + \frac{x^2}{2!} f''(0) + \ldots +$$

$$\frac{x^{n-1}}{(n-1)!} f^{n-1}(0) + \frac{x^n}{n!} f^n(\theta x)$$

where, $\dfrac{x^n}{n!} f^n(\theta x) = R_n$ is the remainder after n terms.

$R_n \to 0$ as $n \to \infty$, because

$$\lim_{n \to \infty} \left| \frac{R_{n+1}}{R_n} \right| = 0 < 1 \; ; \quad \forall \, x \in \mathbb{R}$$

$\therefore \qquad \cos x = 1 - \frac{x^2}{2!} + \frac{x^4}{4!} - \frac{x^6}{6!} + \ldots \; \forall \, x \in \mathbb{R}$

(iv) Let,

$$f(x) = (1 + x)^m$$
$$f'(x) = m \, (1 + x)^{m-1}$$
$$f''(x) = m \, (m - 1) \, (1 + x)^{m-2}$$
$$f'''(x) = m \, (m - 1) \, (m - 2) \, (1 + x)^{m-3}$$
$$\vdots$$
$$f^n(x) = m \, (m - 1) \ldots (m - n + 1) + (1 + x)^{m-n}$$

$\Rightarrow \qquad f(0) = 1, \; f'(0) = m, \; f''(0) = m(m - 1),$
$$f'''(0) = m(m - 1)\,(m - 2)$$

$\therefore$ By Maclaurin's theorem

$$(1 + x)^m = 1 + mx + \frac{m(m - 1)}{2!} x^2 + \frac{m(m - 1)\,(m - 2)}{3!} x^3 + \ldots + R_n$$

where, R_n is the remainder after n terms.

$R_n \to 0$ as $n \to \infty$, because

$$\lim_{n \to \infty} \left| \frac{R_{n+1}}{R_n} \right| = |x| < 1 \qquad\qquad \forall \, x \in \mathbb{R}$$

$\therefore \quad (1 + x)^m = 1 + mx + \frac{m(m - 1)}{2!} x^2 + \ldots \qquad\qquad \forall \, x \in (-1, 1)$

$$(1 + x)^m = 1 + mx + \frac{m(m - 1)}{2!} x^2 + \frac{m(m - 1)\,(m - 2)}{3!} x^3 + \ldots + x^m$$

if m is a positive integer.

(v) Let

$$f(x) = \log (1 + x)$$

$f'(x) = \dfrac{1}{1 + x} \qquad\qquad f(0) = \log 1 = 0$

$f'''(x) = \dfrac{-1}{(1 + x)^2} \qquad\qquad f'(0) = 1$

$f'''(x) = \dfrac{(-1)\,(-2)}{(1 + x)^3} \qquad\qquad f''(0) = -1$

$f^{iv}(x) = \dfrac{(-1)\,(-2)\,(-3)}{(1 + x)^4} \qquad\qquad f'''(0) = -1$

$\qquad\qquad \vdots \qquad\qquad\qquad\qquad f^{iv}(0) = -3!$

$f^n(x) = \dfrac{(-1)^{n-1}\,(n - 1)!}{(1 + x)^n} \qquad$ and so on

By Maclaurin's theorem

$$\log(1+x) = 0 + x\cdot 1 + \frac{x^2}{2!}(-1) + \frac{x^3}{3!}(2!) + \frac{x^4}{4!}(-3!) + \ldots + R_n$$

$R_n \to 0$ as $n \to \infty$, because

$$\lim_{n\to\infty}\left|\frac{R_{n+1}}{R_n}\right| = |x| < 1 \qquad \forall\, x \in (-1, 1)$$

$$\therefore \qquad \log(1+x) = x - \frac{x^2}{2} + \frac{x^3}{3} - \frac{x^4}{4} + \ldots \qquad\qquad \forall\, x \in (-1, 1)$$

Example 2.18 : *Assuming validity of expansions, show that*

$$\frac{1}{2}\,[f(x+h) + f(x-h)] = f(x) + \frac{h^2}{2!}f''(x) + \frac{h^4}{4!}f^{(iv)}(x) + \ldots$$

Solution : By Taylor's series,

$$f(x+h) = f(x) + hf'(x) + \frac{h^2}{2!}f''(x) + \frac{h^3}{3!}f'''(x) + \frac{h^4}{4!}f^{(iv)}(x) + \ldots\ldots \quad (i)$$

$$\therefore\ \ f(x-h) = f(x) - hf'(x) + \frac{h^2}{2!}f''(x) - \frac{h^3}{3!}f'''(x) + \frac{h^4}{4!}f^{(iv)}(x) + \ldots\ldots \quad (ii)$$

$$(i) + (ii) \Rightarrow f(x+h) + f(x-h) = 2f(x) + h^2 f''(x) + \frac{2}{4!}\,h^4 f^{(iv)}(x) + \ldots$$

$$\therefore\ \ \frac{1}{2}\,[f(x+h) + f(x-h)] = f(x) + \frac{h^2}{2!}f''(x) + \frac{h^4}{4!}f^{(iv)}(x) + \ldots\ldots$$

Example 2.19 : *Assuming the validity of expansions, show that*

(i) $\qquad\qquad e^x \cos x = 1 + x - \dfrac{x^3}{3} - \dfrac{x^4}{6} - \dfrac{x^5}{30} + \ldots$

(ii) $\qquad\qquad \log \sec x = \dfrac{x^2}{2!} + \dfrac{2x^4}{4!} + \dfrac{16x^5}{6!} + \ldots$

Solution : (i) By Maclaurin's series expansion, we have

$$f(x) = f(0) + xf'(0) + \frac{x^2}{2!}f''(0) + \frac{x^3}{3!}f'''(0) + \ldots \quad \ldots (i)$$

Let $\qquad f(x) = e^x \cos x, \ f(0) = 1$

$$f^n(x) = 2^{\frac{n}{2}}e^x \cos\left(x + \frac{n\pi}{4}\right)$$

$$\therefore \qquad f'(0) = \sqrt{2}.\cos\frac{\pi}{4} = \sqrt{2}\cdot\frac{1}{\sqrt{2}} = 1$$

$$f''(0) = 2.\cos\left(\frac{\pi}{2}\right) = 0$$

$$f'''(0) = 2\sqrt{2}\cos\left(\frac{3\pi}{4}\right)$$

$$= 2\sqrt{2}\cdot\left(\frac{-1}{2}\right) = -2$$

$$f^{(iv)}(0) = 4\cos(\pi) = -4$$

Using all the above values, we have by (i),

$$e^x \cos x = 1 + x - \frac{x^3}{3} - \frac{x^4}{6} - \frac{x^5}{30} + \cdots$$

(ii) By Maclaurin's series expansion, we have

$$f(x) = f(0) + xf'(0) + \frac{x^2}{2!}f''(0) + \frac{x^3}{3!}f'''(0) + \frac{x^4}{4!}f^{(iv)}(0) + \quad \cdots \text{(i)}$$

$$\text{Let} \quad f(x) = \log \sec x,\ f(0) = 0$$

$$f'(x) = \tan x,\ f'(0) = 0$$

$$f''(x) = \sec^2 x = 1 + \tan^2 x = 1 + [f'(x)]^2 \Rightarrow f''(0) = 1$$

$$f'''(x) = 2f'(x)\cdot f''(x) \Rightarrow f'''(0) = 0$$

$$f^{(iv)}(x) = 2f'(x)f'''(x) + 2[f''(x)]^2 \Rightarrow f^{(iv)}(0) = 2$$

$$f^{(v)}(x) = 2f'(x)f^{(iv)}(x) + 2f''(x)\ f'''(x) + 4f''(x)f'''(x)$$

$$= 2f'(x)f^{(iv)}(x) + 6f''(x)\ f'''(x) \Rightarrow f^{(v)}(0) = 0$$

$$f^{(vi)}(x) = 2f'(x)f^{(v)}(x) + 2f''(x)f^{(iv)}(x) + 6f''(x)\ f^{(iv)}(x)$$

$$+ 6[f'''(x)]^2$$

$$\Rightarrow \quad f^{(vi)}(0) = 2.1.2 + 6.2 = 16$$

Using all the above values, (i) can be written as

$$\log \sec x = \frac{x^2}{2!} + \frac{2x^4}{4!} + \frac{16x^5}{6!} + \cdots$$

Example 2.20 : *Apply Taylor's theorem to prove that*

(i) $\log \sin(x+h) = \log \sin x + h \cot x - \frac{h^2}{2}\cosec^2 x + \frac{h^3}{3}\cot x$ $\cosec^2 x + \cdots$

(ii) $\sin^{-1}(x+h) = \sin^{-1} x + \frac{h}{\sqrt{1-x^2}} + \frac{x}{(1-x)^{3/2}}\frac{h}{2!} + \cdots$

Solution : (i) By Taylor's theorem, we have

$$f(x+h) = f(x) + hf'(x) + \frac{h^2}{2!}f''(x) + \frac{h^3}{3!}f'''(x) + \cdots \ \cdots \text{(i)}$$

$$\text{Let} \quad f(x+h) = \log \sin(x+h)$$

$$\text{Put } h = 0, \quad f(x) = \log \sin x$$

$$\therefore \qquad f'(x) = \cot x$$

$$f''(x) = -\operatorname{cosec}^2 x$$

$$f'''(x) = -2 \operatorname{cosec} x \, (-\operatorname{cosec} x \cot x)$$

$$= 2 \operatorname{cosec}^2 x \cot x, \text{ and so on.}$$

Using all the above values, (i) becomes

$$\log \sin (x + h) = \log \sin x + h \cot x - \frac{h^2}{2}\operatorname{cosec}^2 x$$

$$+ \frac{h^3}{3} \operatorname{cosec}^2 x \cot x + \dots$$

(ii) Let $\qquad f(x + h) = \sin^{-1} (x + h)$

     Put $\qquad h = 0, \ f(x) = \sin^{-1} x$

$$\therefore \qquad f'(x) = \frac{1}{\sqrt{1 - x^2}} = (1 - x)^{-1/2}$$

$$f''(x) = -\frac{1}{2} (1 - x^2)^{3/2} (-2x)$$

$$= \frac{x}{(1 - x^2)^{3/2}} \text{ and so on.}$$

Using all these values in (i), we have

$$\sin^{-1} (x + h) = \sin^{-1} x + \frac{h}{\sqrt{1 - x^2}} + \frac{x}{(1 - x^2)^{3/2}} \cdot \frac{h}{2!} + \dots$$

Example 2.21 : *(i) Expand $\sin x$ in ascending powers of $\left(x - \dfrac{\pi}{2}\right)$.*

(i) Expand $\log \sin x$ in ascending powers of $(x - 3)$.

(ii) Expand $\tan x$ in ascending powers of $\left(x - \dfrac{\pi}{4}\right)$.

Solution : (i) By Taylor's theorem, we have

$$f(a + h) = f(a) + hf'(a) + \frac{h^2}{2!}f''(a) + \frac{h^3}{3!}f'''(a) \qquad \dots(i)$$

We take, $\qquad h = x - \dfrac{\pi}{2}, \ a = \dfrac{\pi}{2}$

$$\therefore \qquad f(a + h) = f(x)$$

Let $\qquad f(x) = \sin x$

$\therefore$ The result (i) reduces to

$$\sin x = f\left(\frac{\pi}{2}\right) + \left(x - \frac{\pi}{2}\right) f'\left(\frac{\pi}{2}\right) + \frac{\left(x - \frac{\pi}{2}\right)^2}{2!} f''\left(\frac{\pi}{2}\right) + \frac{\left(x - \frac{\pi}{2}\right)^3}{3!} f'''\left(\frac{\pi}{2}\right) + \dots$$

$$\dots (ii)$$

Now, $\qquad f(x) = \sin x \Rightarrow f\left(\dfrac{\pi}{2}\right) = \sin \dfrac{\pi}{2} = 1$

$$f'(x) = \cos x \Rightarrow f'\left(\dfrac{\pi}{2}\right) = \cos \dfrac{\pi}{2} = 0$$

$$f''(x) = -\sin x \Rightarrow f''\left(\dfrac{\pi}{2}\right) = -\sin \dfrac{\pi}{2} = -1$$

$$f'''(x) = -\cos x \Rightarrow f'''\left(\dfrac{\pi}{2}\right) = -\cos \dfrac{\pi}{2} = 0$$

$$f^{(iv)}(x) = \sin x \Rightarrow f^{(iv)}\left(\dfrac{\pi}{2}\right) = \sin \dfrac{\pi}{2} = 1 \text{ and so on.}$$

Substituting these values in (ii),

$$\sin x = 1 - \frac{1}{2!}\left(x - \frac{\pi}{2}\right)^2 + \frac{1}{4!}\left(x - \frac{\pi}{2}\right)^4 - \frac{1}{6!}\left(x - \frac{\pi}{2}\right)^6 + \ldots$$

(ii) By Taylor's theorem, we have

$$f(a + h) = f(a) + hf'(a) + \frac{h^2}{2!}f''(a) + \frac{h^3}{3!}f'''(a) + \ldots \quad (i)$$

We take $\qquad h = x - 3, \ a = 3$

$\therefore \qquad\qquad f(a + h) = f(x)$

Let $\qquad\qquad f(x) = \log \sin x$

$\therefore$ The result (i) reduces to

$$\log \sin x = f(3) + (x - 3)\,f'(3) + \frac{(x - 3)^2}{2!}f''(3) + \frac{(x - 3)^3}{3!}f'''(3) + \ldots$$

$$\ldots (ii)$$

Now, $\qquad f(x) = \log \sin x \Rightarrow f(3) = \log \sin 3$

$$f'(x) = \frac{\cos x}{\sin x} \Rightarrow f'(3) = \cot 3$$

$$f''(x) = -\operatorname{cosec}^2 x \Rightarrow f''(3) = -\operatorname{cosec}^2 3$$

$$f'''(x) = -2\operatorname{cosec} x\,(-\operatorname{cosec} x \cot x) = 2\operatorname{cosec}^2 x \cot x$$

$$\Rightarrow f'''(3) = 2\operatorname{cosec}^2 3 \cot 3 \text{ and so on.}$$

Substituting these values in (ii),

$$\log \sin x = \log \sin 3 + (x - 3)\cot 3 - \frac{(x - 2)^2}{2}\operatorname{cosec}^2 3 + \frac{(x - 3)^3}{3}$$

$\operatorname{cosec}^2 3 \cot 3 + \ldots$

(iii) By Taylor's theorem, we have

$$f(a + h) = f(a) + hf'(a) + \frac{h^2}{2!}f''(a) + \frac{h^3}{3!}f'''(a) + \ldots \qquad \ldots (i)$$

We take $\qquad h = x - \dfrac{\pi}{4},\ a = \dfrac{\pi}{4}$

$\therefore \qquad f(a + h) = f(x)$

$\therefore$ The result (i) reduces to

$$\tan x = f\!\left(\frac{\pi}{4}\right) + \left(x - \frac{\pi}{4}\right) f'\!\left(\frac{\pi}{4}\right) + \frac{\left(x - \frac{\pi}{4}\right)^2}{2!} f''\!\left(\frac{\pi}{4}\right) + \frac{\left(x - \frac{\pi}{4}\right)^3}{3!} f'''\!\left(\frac{\pi}{4}\right) + \ldots$$

$$\ldots \text{(ii)}$$

Let, $\qquad f(x) = \tan x \Rightarrow f\!\left(\dfrac{\pi}{4}\right) = 1$

$$f'(x) = \sec^2 x \Rightarrow f'\!\left(\frac{\pi}{4}\right) = \left(\sqrt{2}\right)^2 = 1$$

$$f''(x) = 2 \sec x \ \sec x \ \tan x = 2 \sec^2 x \tan x$$

$$= f''\!\left(\frac{\pi}{4}\right) = 2\left(\sqrt{2}\right)^2 . 1 = 4$$

$$f'''(x) = 2(2 \sec x \sec x \tan x . \tan x + \sec^2 x . \sec^2 x)$$

$$= 2 \sec^2 x \, (2 \tan^2 x + \sec^2 x)$$

$$= f'''\!\left(\frac{\pi}{4}\right) = 2(2) \ (2 + 2) = 16 \qquad \text{and so on.}$$

Substituting these values in (ii),

$$\tan x = 1 + 2\left(x - \frac{\pi}{4}\right) + 2\left(x - \frac{\pi}{4}\right)^2 + \frac{8}{3}\left(x - \frac{\pi}{4}\right)^3 + \ldots$$

Example 2.22 : *Expand (i) tan x, (ii) $\log \sqrt{\dfrac{1+x}{1-x}}$.*

Solution : (i) Let $y = f(x) = \tan x \Rightarrow y(0) = 0$

$y_1 = \sec^2 x = 1 + \tan^2 x = 1 + y^2 \Rightarrow y_1(0) = 1 + 0 = 1$

$y_2 = 2y\, y_1 \Rightarrow y_2(0) = 0$

$y_3 = 2y_1\, y_1 + 2y\, y_2 = 2y_1^2 + 2yy_2 \Rightarrow y_3(0) = 2 + 0 = 2$

$y_4 = 4y_1 y_2 + 2yy_3 + 2y_1 y_2 = 6y_1 y_2 + 2yy_3 \Rightarrow y_4(0) = 6(1)\,0 + 0 = 0$

$y_5 = 8y_1 y_3 + 6y_2^2 + 2yy_4 \Rightarrow y_5(0) = 8(1)\,(2) + 0 = 16$

$$\therefore \tan x = y(0) + xy_1(0) + \frac{x^2}{2!} y_2(0) + \frac{x^3}{3!} y_3(0) + \frac{x^4}{4!} y_4(0) + \frac{x^5}{5!} y_5(0) + \ldots$$

$$\tan x = x + \frac{x^3}{3!}\,(2) + \frac{x^5}{5!}\,(16) + \ldots$$

$$= x + \frac{x^3}{3} + \frac{2x^5}{15} + \ldots$$

(ii) We know that,

$$\log (1 + x) = x - \frac{x^2}{2} + \frac{x^3}{3} - \frac{x^4}{4} + \ldots$$

$$\text{and } \log (1 - x) = -x - \frac{x^2}{2} - \frac{x^3}{3} - \frac{x^4}{4} - \ldots$$

$$\therefore \quad \log \sqrt{\frac{1 + x}{1 - x}} = \frac{1}{2} [\log (1 + x) - \log (1 - x)]$$

$$= \frac{1}{2} \left[\left(x - \frac{x^2}{2} + \frac{x^3}{3} - \frac{x^4}{4} + \ldots \right) - \left(-x - \frac{x^2}{2} - \frac{x^3}{3} - \frac{x^4}{4} - \ldots \right) \right]$$

$$= \frac{1}{2} \left[2 \left(x + \frac{x^3}{3} + \frac{x^5}{5} - \frac{x^7}{7} + \ldots \right) \right]$$

$$\Rightarrow \quad \log \sqrt{\frac{1 + x}{1 - x}} = x + \frac{x^3}{3} + \frac{x^5}{5} - \frac{x^7}{7} + \ldots$$

Example 2.23 : *Prove that*

$$tan^{-1} x = x - \frac{1}{3} x^3 + \frac{1}{5} x^5 \ldots \text{ and hence find the value of } \pi.$$

Solution : Let $y = \tan^{-1} x \Rightarrow y (0) = \tan^{-1} 0 = 0$

$$y_1 = \frac{1}{1 + x^2} = (1 + x^2)^{-1}$$

$$= 1 - x^2 + x^4 - x^6 + \ldots \qquad \Rightarrow y_1 (0) = 1$$

$$y_2 = -2x + 4x^3 - 6x^5 + \ldots \qquad \Rightarrow y_2 (0) = 0$$

$$y_3 = -2 + 12x^2 - 30x^4 + \ldots \qquad \Rightarrow y_3 (0) = -2$$

$$y_4 = 24x - 120x^3 + \ldots \qquad \Rightarrow y_4 (0) = 0$$

$$y_5 = 24 - 360x^2 + \ldots \qquad \Rightarrow y_5 (0) = 24$$

$$\therefore \quad \tan^{-1} x = y(0) + xy_1(0) + \frac{x^2}{2!} y_2(0) + \frac{x^3}{3!} y_3 (0) + \ldots$$

$$\Rightarrow \quad \tan^{-1} x = x - \frac{2x^3}{3!} + \frac{24x^5}{5!} - \ldots$$

$$\Rightarrow \quad \tan^{-1} x = x - \frac{x^3}{3} + \frac{x^5}{5} - \frac{x^7}{7} + \ldots$$

Putting $x = 1$, we have

$$\tan^{-1} 1 = 1 - \frac{1}{3} + \frac{1}{5} - \frac{1}{7}$$

$$\therefore \quad \frac{\pi}{4} = 1 - \frac{1}{3} + \frac{1}{5} - \frac{1}{7} + \ldots$$

$$\therefore \quad \pi = 4 \left[1 - \frac{1}{3} + \frac{1}{5} - \frac{1}{7} + \ldots \right]$$

Example 2.24 : *Prove that*

(i) $\sin^{-1} x = x + 1^2 \cdot \dfrac{x^3}{3!} + 1^2 \cdot 3^2 \cdot \dfrac{x^5}{5!} + 1^2 \cdot 3^2 \cdot 5^2 \cdot \dfrac{x^7}{7!} + \ldots$

(ii) $\cos (m \sin^{-1} x) = 1 - \dfrac{m^2 x^2}{2!} + \dfrac{m^2(m^2 - 2^2)}{4!} x^4$

$$- \dfrac{m^2 (m^2 - 2^2)(m^2 - 4^2)}{6!} x^6 + \ldots$$

Solution : (i) Let $y = \sin^{-1} x \Rightarrow y(0) = \sin^{-1}(0) = 0$

$$y_1 = \dfrac{1}{\sqrt{1 - x^2}} \Rightarrow y_1(0) = 1$$

$$y_1 = (1 - x^2)^{-1/2}$$

$$= 1 + \dfrac{x^2}{2} + \dfrac{\frac{1}{2}\left(\frac{1}{2} + 1\right)}{2!} x^4 + \dfrac{\frac{1}{2}\left(\frac{1}{2} + 1\right)\left(\frac{1}{2} + 2\right)}{3!} x^6 + \ldots \ldots$$

$$= 1 + \dfrac{1}{2} x^2 + \dfrac{3}{8} x^4 + \dfrac{5}{16} x^6 + \ldots$$

$$y_2 = 1 + x + \dfrac{3}{2} x^3 + \dfrac{15}{8} x^5 + \ldots \qquad\qquad \Rightarrow y_2(0) = 0$$

$$y_3 = 1 + \dfrac{9}{2} x^2 + \dfrac{75}{8} x^4 + \ldots \qquad\qquad \Rightarrow y_3(0) = 1$$

$$y_4 = 9x + \dfrac{75}{2} x^3 + \ldots \qquad\qquad \Rightarrow y_4(0) = 0$$

$$y_5 = 9 + \dfrac{225}{2} x^2 + \ldots \qquad\qquad \Rightarrow y_5(0) = 9$$

$$y_6 = 225 x + \ldots \qquad\qquad \Rightarrow y_6(0) = 0$$

$$y_7 = 225 \qquad\qquad \Rightarrow y_7(0) = 225$$

$$\therefore \quad \sin^{-1} x = y(0) + xy_1(0) + \dfrac{x^2}{2!} y_2(0) + \dfrac{x^3}{3!} y_3(0) + \dfrac{x^4}{4!} y_4(0) + \ldots$$

$$= x + \dfrac{x^3}{3!} + \dfrac{9 x^5}{5!} + \dfrac{225 x^7}{7!} + \ldots$$

$$\Rightarrow \sin^{-1} x = x + 1^2 \cdot \dfrac{x^3}{3!} + 1^2 \cdot 3^2 \cdot \dfrac{x^5}{5!} + 1^2 \cdot 3^2 \cdot 5^2 \dfrac{x^7}{7!} + \ldots$$

(ii) $y = \cos (m \sin^{-1} x) \Rightarrow y(0) = \cos 0 = 1$

$$y_1 = - \sin (m \sin^{-1} x) \cdot \dfrac{m}{\sqrt{1 - x^2}} \Rightarrow y_1(0) = 0$$

Squaring and cross-multiplying,

$$(1 - x^2)\, y_1^2 = m^2 \sin^2 (m \sin^{-1} x)$$

$$\Rightarrow \qquad (1 - x^2)\, y_1^2 = m^2 [1 - \cos^2 (m \sin^{-1} x)]$$

$$= m^2 (1 - y^2)$$

Differentiating this, we get,

$$2(1 - x^2)\, y_1 y_2 - 2x y_1^2 = m^2 (-2y y_1)$$

$$\Rightarrow (1 - x^2)\, y_2 - x y_1 + m^2 y = 0$$

$$\Rightarrow \qquad\qquad y_2(0) = -m^2$$

Differentiating n times by Leibnitz's rule, we get

$$(1 - x^2)\, y_{n+2} - (2n + 1)\, x y_{n-1}\, (n^2 - m^2)\, y_n = 0$$

Put $x = 0$, $\quad y_{n+2}(0) = (n^2 - m^2)\, y_n(0)$

Put $\qquad\qquad n = 1, 2, 3 \ldots\ldots$

$$y_3(0) = (1^2 - m^2)\, y_1(0) = 0$$

$$y_4(0) = (2^2 - m^2)\, y_2(0) = -m^2 (2^2 - m^2)$$

$$y_5(0) = 0$$

$$y_6(0) = (4^2 - m^2)\, y_4(0) = -m^2 (2^2 - m^2)(4^2 - m^2)$$

and so on.

When n is odd, $y_n(0) = 0$

When n is even, $y_n(0) = -m^2 (2^2 - m^2)(4^2 - m^2)\ldots[(n-2)^2 - m^2]$

Also, $\cos (m \sin^{-1} x) = y(0) + x y_1(0) + \dfrac{x^2}{2!} y_2(0) + \dfrac{x^3}{3!} y_3(0) + \ldots$

$$= 1 - \dfrac{m^2 x^2}{2!} + \dfrac{m^2 (2^2 - m^2)}{4!} x^4 - \dfrac{m^2 (2^2 - m^2)(4^2 - m^2)}{6!} x^6 + \ldots$$

Example 2.25 : *Obtain the expansion of $e^{\sin x}$ upto the first four terms.*

Solution : Let $\quad f(x) = e^{\sin x} \Rightarrow f(0) = e^0 = 1$

$$f'(x) = e^{\sin x} . \cos x = f(x) . \cos x \Rightarrow f'(0)$$

$$= 1.1 = 1$$

$$f''(x) = f'(x) \cos x - f(x) \sin x \Rightarrow f'(0) = 1$$

$$f'''(x) = f''(x) \cos x - f'(x) \sin x - f'(x) \sin x$$

$$- f(x) \cos x$$

$$= f''(x) \cos x - 2f'(x) \sin x - f(x) \cos x$$

$$\Rightarrow \qquad f'''(0) = 0$$

$$f^{(iv)}(x) = f'''(x)\,\cos x - f''(x)\,\sin x - 2f''(x)\,\sin x$$
$$- 2f'(x)\,\cos x - f'(x)\,\cos x + f(x)\,\sin x$$

$$\Rightarrow \qquad f^{(iv)}(0) = -2 - 1 = -3$$

By Maclaurin's series expansion,

$$e^{\sin x} = f(0) + x\,f'(0) + \frac{x^2}{2!}f''(0) + \frac{x^3}{3!}f'''(0)$$
$$+ \frac{x^4}{4!}f^{(iv)}(0) + \dots$$

$$\therefore \qquad e^{\sin x} = 1 + x + \frac{x^2}{2} - \frac{x^4}{8} + \dots$$

Example 2.26 : *Expand cosh x and sinh x by Maclaurin's series.*

Solution : We have,

$$\cosh x = \frac{e^x + e^{-x}}{2}, \quad x \in R$$

We know that, $\qquad e^x = 1 + x + \dfrac{x^2}{2!} + \dfrac{x^3}{3!} + \dfrac{x^4}{4!} + \dots$

and $\qquad\qquad\qquad e^{-x} = 1 - x + \dfrac{x^2}{2!} - \dfrac{x^3}{3!} + \dfrac{x^4}{4!} - \dots$

$$\therefore \qquad \cosh x = \frac{e^x + e^{-x}}{2}$$

$$= \frac{1}{2}\left[\left(1 + x + \frac{x^2}{2!} + \frac{x^3}{3!} + \frac{x^4}{4!} + \dots\right) + \left(1 - x + \frac{x^2}{2!} - \frac{x^3}{3!} + \frac{x^4}{4!} - \dots\right)\right]$$

$$\therefore \qquad \cosh x = 1 + \frac{x^2}{2!} + \frac{x^4}{4!} + \frac{x^6}{6!} + \dots$$

and $\qquad\qquad\qquad \sinh x = \dfrac{e^x - e^{-x}}{2}$

$$\therefore \qquad \sinh x = x + \frac{x^3}{3!} + \frac{x^5}{5!} + \frac{x^7}{7!} + \dots$$

Example 2.27 : *Using Taylor's theorem, evaluate*

(i) $\displaystyle \lim_{x \to 0} \frac{2(\tan x - \sin x) - x^3}{x^5}$

(ii) $\displaystyle \lim_{x \to 0} \left[\frac{\log\left(\dfrac{1+x}{1-x}\right) - 2x}{x^5}\right]$

Solution : (i) $\displaystyle \lim_{x \to 0} \frac{2\,(\tan x - \sin x) - x^3}{x^5}$

$$= \lim_{x \to 0} \frac{2\left[\left(x + \dfrac{x^3}{3!} + \dfrac{2x^5}{15} + \ldots\right) - \left(x - \dfrac{x^2}{6} - \dfrac{x^5}{120} \cdots\right)\right] - x^3}{x^5}$$

$$= \lim_{x \to 0} \frac{2\left(\dfrac{1}{2}x^3 + \dfrac{1}{8}x^5 + \ldots\right) - x^3}{x^5}$$

$$= \lim_{x \to 0} \frac{x^3 + \dfrac{1}{4}x^5 + \ldots - x^3}{x^5}$$

$$= \lim_{x \to 0} \left(\frac{1}{4} + \text{terms containing powers of x}\right)$$

$$= \frac{1}{4} + 0 + 0$$

$$= \frac{1}{4}$$

(ii) $\displaystyle \log\left(\frac{1+x}{1-x}\right) = \log(1+x) - \log(1-x)$

$$= \left(x - \frac{x^2}{2} + \frac{x^3}{3} - \frac{x^4}{4}\right) - \left(-x - \frac{x^2}{2} - \frac{x^3}{3} - \frac{x^4}{4} \cdots\right)$$

$$= 2x + \frac{2}{3}x^3 + \frac{2}{5}x^5 + \ldots$$

$$\therefore \quad \lim_{x \to 0} \frac{\log\dfrac{1+x}{1-x} - 2x}{x^3} = \lim_{x \to 0} \frac{2x + \dfrac{2}{3}x^3 + \dfrac{2}{5}x^5 + \ldots - 2x}{x^3}$$

$$= \lim_{x \to 0} \frac{\dfrac{2}{3}x^3 + \dfrac{2}{5}x^5 + \ldots}{x^3}$$

$$= \lim_{x \to 0} \left(\frac{2}{3} + \frac{2}{5}x^2 + \text{terms containing powers of x}\right)$$

$$= \frac{2}{3} + 0 + 0$$

$$= \frac{2}{3}$$

Think Over It

1. Using L' Hospital's Rule, prove that the series $\displaystyle\sum_{n=1}^{\infty} \frac{(-1)^n\, n^3}{(n^2 + 1)^{4/3}}$ diverges.

2. Using L' Hospital's Rule, show that the series $\displaystyle\sum_{n=1}^{\infty} n\, \sin\left(\frac{1}{n}\right)$ is divergent.

3. Let f be a function such that $f^{(n)}_{(a)}$ exists. Then prove that,
$$f(a + y) = f(a) + hf'(a) + \frac{h^2}{2!} f''(a) + \ldots + \frac{h^n}{n!} f^{(n)}_{(a)} + 0(h^n).$$

4. When can we express the function as an infinite series in Taylor's theorem ?

Summary

1. $\dfrac{0}{0}, \dfrac{\infty}{\infty}, \infty - \infty, 0 \cdot \infty, 1^{\infty}, 0^0, \infty^0$ are all indeterminate forms.

2. (i) If $\quad\quad y = (ax + b)^m$, m is a positive integer

then $\quad y_n = \dfrac{m!}{(m - n)!} a^n (ax + b)^{m-n}$; if $n < m$

$\quad\quad\quad\quad = m!\, a^m \quad\quad\quad\quad$; if $n = m$

$\quad\quad\quad\quad = 0 \quad\quad\quad\quad\quad\quad$; if $n > m$

(ii) If $\quad\quad y = \dfrac{1}{ax + b}$

then $\quad y_n = \dfrac{(-1)^n\, n!\, a^n}{(ax + b)^{n+1}}$

(iii) If $\quad\quad y = \log(ax + b)$

then $\quad y_n = \dfrac{(-1)^{n-1}\, (n - 1)!\, a^n}{(ax + b)^n}$

(iv)　If　　　　　　$y = e^{mx}$

　　　then　　　　　$y_n = m^n e^{mx}$

(v)　If　　　　　　$y = a^{mx}$

　　　then　　　　　$y_n = m^n a^{mx} (\log a)^n$

(vi)　If　　　　　　$y = \sin(ax + b)$

　　　then　　　　　$y_n = a^n \sin\left(ax + b + \dfrac{n\pi}{2}\right)$

(vii)　If　　　　　$y = \cos(ax + b)$

　　　then　　　　　$y_n = a^n \cos\left(ax + b + \dfrac{n\pi}{2}\right)$

(viii)　If　　　　　$y = e^{ax} \sin(bx + c)$

　　　then　　　　　$y_n = r^n e^{ax} \sin(bx + c + n\theta)$

(ix)　If　　　　　　$y = e^{ax} \cos(bx + c)$

　　　then　　　　　$y_n = r^n e^{ax} \cos(bx + c + n\theta)$

$$\left(\text{where, } r = (a^2 + b^2)^{1/2} \text{ and } \theta = \tan^{-1}\frac{b}{a}\right)$$

3.　**Leibnitz's theorem**

If $y = uv$ where u and v are functions of x possessing derivatives of nth order then

$$y_n = (uv)_n = n_{c_0} u_n v + n_{c_1} u_{n-1} v_1 + n_{c_2} u_{n-2} v_2 + \dots + n_{c_r} u_{n-r} v_r + \dots + n_{c_n} u v_n$$

where, $n_{c_r} = \dfrac{n!}{(n-r)!\, r!}$

4.　Taylor's theorem with Lagrange's form of remainder :

(i)　f and its derivatives $f', f'' \dots f^{n-1}$ are continuous in [a, b].

(ii)　f^n exists in (a, b) then $\exists\, c \in (a,b)$ such that

$$f(b) = f(a) + (b-a)\, f'(a) + \frac{(b-a)^2}{2!} f''(a) + \dots + \frac{(b-a)^{n-1}}{(n-1)!} f^{n-1}(a)$$

$$+ \frac{(b-a)^n}{n!} f^n(c) \qquad\qquad \dots (i)$$

where, $\dfrac{(b-a)^n}{n!} f^n(c) = R_n$ is called Lagrange's form of remainder.

Note :

1. In above theorem, if we consider the interval as $[a, a + h]$ then above theorem becomes

$$f(a + h) = f(a) + h\, f'(a) + \frac{h^2}{2!} f''(a) + \ldots + \frac{h^n}{n!} f^n(a + \theta h) \quad \ldots \text{(ii)}$$

where, $\dfrac{h^n}{n!} f^n(a + \theta h) = R_n$ is Lagrange's theorem of remainder

2. In above results equation (ii). If we put $a = 0$ and $h = x$, we get Maclaurin's theorem.

$$f(x) = f(0) = x\, f'(a) + \frac{x^2}{2!} f''(a) + \ldots + \frac{x^n}{n!} f^n(c) \quad \text{where } 0 < c < x$$

$$\ldots \text{(iii)}$$

5. $2 \sin A \sin B = \cos (A - B) - \cos (A + B)$

$2 \sin A \cos B = \sin (A + B) + \sin (A - B)$

$2 \cos A \cos B = \cos (A + B) + \cos (A - B)$

$2 \cos A \sin B = \sin (A + B) - \sin (A - B)$

Exercise

[A] Say True or False : Justify !

1. $\lim\limits_{x \to a} f(x) = 0$ and $\lim\limits_{x \to a} g(x) = 1$, then $\lim\limits_{x \to a} \dfrac{f(x)}{g(x)} = \lim\limits_{x \to a} \dfrac{f'(x)}{g'(x)}$.

2. If $\lim\limits_{x \to a} f(x) = 10$ and $\lim\limits_{x \to a} g(x) = 0$, then $\lim\limits_{x \to a} \dfrac{f(x)}{g(x)} = \lim\limits_{x \to a} \dfrac{f'(x)}{g'(x)}$.

3. If $\lim\limits_{x \to a} f(x) = 0$ and $\lim\limits_{x \to a} g(x) = 0$, then $\lim\limits_{x \to a} \dfrac{f(x)}{g(x)} = \lim\limits_{x \to a} \dfrac{f'''(x)}{g'''(x)}$.

4. Every function can be expressed as an infinite series.

5. If $y = \dfrac{1}{ax + b}$, then $y_n = \dfrac{(-1)^n\, n!\, a_n}{(ax + b)^n}$.

6. If $y = (ax + b)^n$, then there exists $m \in \mathbb{N}$, such that

$$y_k(x) = 0 \ \forall \ k \geq m$$

[B] Multiple Choice Questions : Choose the Correct Alternative

1. Which of the following is an indeterminate form ......

 (a) ∞^0 (b) $\infty + \infty$

 (c) 1^0 (d) none of these

2. $\displaystyle \lim_{x \to 1} (1 - x) \tan \frac{\pi x}{2}$ is ......

(a) 0

(b) $\dfrac{\pi}{2}$

(c) $\dfrac{2}{\pi}$

(d) none of these

3. $\displaystyle \lim_{x \to 0} x^x$ is ......

(a) 0

(b) e

(c) 1

(d) none of these

4. $\displaystyle \lim_{x \to \infty} (a^{1/x} - 1) x$ is ......

(a) a

(b) log a

(c) 0

(d) none of these

5. $\displaystyle \lim_{x \to 0} \frac{a^x - 1}{b^x - 1}$ is ......

(a) log ab

(b) log a

(c) log b

(d) none of these

6. If $e^x = a_0 + a_1 x + a_2 x^2 + \ldots$ then value of $a_3 = $ ......

(a) 0

(b) 1

(c) $\dfrac{1}{3!}$

(d) none of these

7. If $\sin x = a_0 + a_1 x + a_2 x^2 + a_3 x^3 + \ldots$ then the value of $a_2 = $ ......

(a) 0

(b) 1

(c) 2

(d) none of these

8. If $\cos x = a_0 + a_1 x + a_2 x^2 + a_3 x^3 + \ldots$ then the value of $a_0 = $ ......

(1) 0

(b) 1

(c) 2

(d) none of these

9. $f(x) = \tan x$ can be expanded in powers of $\left(x - \dfrac{\pi}{4}\right)$ by using ......

(a) Maclaurin's theorem

(b) Leibnitz's theorem

(c) Taylor's theorem

(d) none of these

10. If $\log (1 + x) = a_0 + a_1 x + a_2 \dfrac{x^2}{2} + a_3 \dfrac{x^3}{3} + \ldots$ then the value of $a_3 = $

(a) 0

(b) 1

(c) 2

(d) none of these

11. If $y = x^m$, m is a positive integer, then $y_n = 0$ if ......

    (a) $n < m$ (b) $n = m$

    (c) $n > m$ (d) none of these

12. If $y = \sin(ax + b)$, then $y_n = $ ......

    (a) $a \sin\left(ax + b + \dfrac{n\pi}{2}\right)$ (b) $a^n \sin\left(ax + b + \dfrac{(n-1)\pi}{2}\right)$

    (c) $a^n \sin\left(ax + b + \dfrac{n\pi}{2}\right)$ (d) none of these

13. If $y = \sin(m \sin^{-1} x)$, then ......

    (a) $(1 - x^2)\, y_2 - xy_1 - m^2 y = 0$ (b) $(1 - x^2)\, y_2 - xy_1 + m^2 y = 0$

    (c) $(1 - x^2) + xy_1 + m^2 y = 0$ (d) none of these

14. If $y = e^{ax + b}$, then $D^n y = $ ......

    (a) $ae^{ax + b}$ (b) $a^n e^{ax + b}$

    (c) $a^{n-1} e^{ax + b}$ (b) none of these

15. If $y = \dfrac{1}{ax + b}$, then $D^n y = $ ......

    (a) $\dfrac{(-1)^n\, n!\, a^n}{(ax + b)^n}$ (b) $\dfrac{(-1)^n\, n!\, a^n}{(ax + b)^{n+1}}$

    (c) $\dfrac{(-1)^{n-1}\, (n-1)!\, a^n}{(ax + b)^n}$ (d) none of these

---

[C] Theory Questions :

1. State L'Hospital's rule for limit of real valued function of a real variable.

2. State and prove the Leibnitz theorem for n^{th} derivative of the product of two differentiable function.

3. State the Maclaurine's theorem for function of real variable.

4. State the Taylor's theorem for a real valued functions of a real variable with Lagrange's form of remainder.

5. If $y = e^{ax} \sin(bx + c)$, then prove that

$$y_n = (a^2 + b^2)^{n/2}\, e^{ax} \sin(bx + c \neq n \tan^{-1}(b/a)).$$

[D] Numerical Problems :

1. Evaluate the following :

(a) $\displaystyle\lim_{x\to 0} \frac{xe^x - \log(1+x)}{x^2}$

(b) $\displaystyle\lim_{x\to 0} \frac{\cosh x \; \cos x}{x \sin x}$

(c) $\displaystyle\lim_{x\to 0} \frac{x \cos x - \log(1+x)}{x^2}$

(d) $\displaystyle\lim_{x\to \frac{1}{2}} \frac{\cos^2 \pi x}{e^{2x} - 2ex}$

(e) $\displaystyle\lim_{x\to 0} \frac{a^x - 1}{b^x - 1}$

(f) $\displaystyle\lim_{x\to 0} \frac{e^x - e^{-x} - 2\log(1+x)}{x \sin x}$

(g) $\displaystyle\lim_{x\to 0} \frac{\log(1-x^2)}{\log \cos x}$

(h) $\displaystyle\lim_{x\to 0} \frac{\sin 2x + 2\sin^2 x - 2\sin x}{\cos x - \cos^2 x}$

(i) $\displaystyle\lim_{x\to a} \frac{\log(x-a)}{\log(e^x - e^a)}$

(j) $\displaystyle\lim_{x\to 0} \log_{\tan x} \tan 2x$

(k) $\displaystyle\lim_{x\to \frac{\pi}{2}} \frac{\tan 5x}{\tan x}$

(l) $\displaystyle\lim_{x\to \pi/2} \frac{\log(x - \pi/2)}{\tan x}$

(m) $\displaystyle\lim_{x\to 0} \left(\frac{1}{x} - \frac{1}{e^x - 1} \right)$

(n) $\displaystyle\lim_{x\to 0} \left(\frac{1}{x^2} - \cot^2 x \right)$

(o) $\displaystyle\lim_{x\to 2} \left(\frac{1}{x-2} - \frac{1}{\log(x-1)} \right)$

(p) $\displaystyle\lim_{x\to \pi/2} (\sec x - \tan x)$

(q) $\displaystyle\lim_{x\to 0} x \log x$

(r) $\displaystyle\lim_{x\to a} (a-x) \tan \frac{\pi x}{2a}$

(s) $\displaystyle\lim_{x\to 0} (\cos x)^{\frac{1}{x^2}}$

(t) $\displaystyle\lim_{x\to a} (x-a)^{x-a}$

(u) $\displaystyle\lim_{x\to \infty} (a^{1/x} - 1) x$

(v) $\displaystyle\lim_{x\to \frac{\pi}{2}} (\sin x)^{\tan x}$

(w) $\displaystyle\lim_{x\to 0} \left(\frac{2x+1}{x+1} \right)^{\frac{1}{x}}$

(x) $\displaystyle\lim_{x\to \frac{\pi}{4}} (\tan x)^{\tan 2x}$

(y) $\displaystyle\lim_{x\to 0} (\operatorname{cosec} x)^{1/\log x}$

2. Find the values of a and b, if $\displaystyle\lim_{x \to 0} \dfrac{x(1 + a \cos x) - b \sin x}{x^3} = 1$

3. If $\displaystyle\lim_{x \to 0} \dfrac{\sin 2x + a \sin x}{x^3}$ is finite, find the limit and the value of a.

4. Evaluate the following limits :

(i) $\displaystyle\lim_{x \to 1} \dfrac{1 + \log x - x}{1 - 2x + x^2}$

(ii) $\displaystyle\lim_{x \to \infty} \dfrac{e^x (1 + x)}{x^2}$

(iii) $\displaystyle\lim_{x \to \infty} \dfrac{x^n}{e^x}$; n is positive integer

(iv) $\displaystyle\lim_{x \to 0} \left(\dfrac{1}{x} - \dfrac{1}{\sin x}\right)$

(v) $\displaystyle\lim_{x \to 0} \left(\dfrac{1}{x^2} - \dfrac{1}{\sin^2 x}\right)$

(vi) $\displaystyle\lim_{x \to 0} \dfrac{1 + \sin x - \cos x + \log (1 - x)}{x \tan^2 x}$

(vii) $\displaystyle\lim_{x \to 0} \left(\dfrac{a^x + b^x}{2}\right)^{1/x}$

(viii) $\displaystyle\lim_{x \to 0} (\cos ax)^{b/x^2}$

(ix) $\displaystyle\lim_{x \to 0} (\cot x)^{\sin 2x}$

(x) $\displaystyle\lim_{x \to \pi/2} (\sin x)^{2 \tan x}$

5. Find the nth derivatives, if

(i) $y = \dfrac{x^4}{(x - 1)(x - 2)}$

(ii) $y = \dfrac{x^2}{(x + 2)(2x + 3)}$

(iii) $y = \dfrac{x + 1}{x^2 - 4}$

(iv) $y = \cos^2 x \, \sin x$

(v) $y = \cos x \cos 2x \cos 3x$

(vi) $y = x^n e^x$

(vii) $y = e^x \log x$

(viii) $y = x^2 e^x \cos x$

(ix) $y = \cos^4 x$

6. If $y = \dfrac{\sin^{-1} x}{\sqrt{1 - x^2}}$ then show that

$$(1 - x^2)\, y_{n+2} - (2n + 3)\, x\, y_{n+1} - (n + 1)^2\, y_n = 0$$

7. If $y = e^{\tan^{-1} x}$, prove that

$$(1 + x^2)\, y_{n+2} + [2(n + 1)\, x - 1]\, y_{n+1} + n(n + 1)\, y_n = 0$$

8. If $y = \tan^{-1} x$, prove that

$$(1 + x^2)\, y_{n+2} + 2(n + 1)\, x\, y_{n+1} + n(n + 1)\, y_n = 0$$

9. If $y = \sin^{-1} x$, show that $(1 - x^2)\, y_{n+2} - 2(n + 1)\, x\, y_{n+1} - n^2 y_n = 0$

10. If $y = a \cos (\log x) + b \sin (\log x)$, show that,

$$x^2 y_{n+2} + (2n + 1)\, x y_{n+1} + (n^2 + 1)\, y_n = 0$$

11. If $y = \log \left(x + \sqrt{a^2 + x^2}\right)^2$, prove that

$$(a^2 + x^2)\, y_{n+2} + (2n + 1)\, x y_{n+1} + n^2\, y_n = 0$$

12. Differentiate the equation

$$x^2 \frac{d^2y}{dx^2} + x \frac{dy}{dx} + (a^2 - m^2)\, y = 0, \text{ n times with respect to x.}$$

13. By obtaining in two different ways the nth derivatives of x^{2n}, prove that

$$1 + \frac{n^2}{1!} + \frac{n^2 (n - 1)^2}{1^2 \cdot 2^2} + \frac{n^2 (n - 1)^2 (n - 2)^2}{1^2 \cdot 2^2 \cdot 3^2} + \cdots = \frac{(2n)!}{(n!)^2}$$

14. Apply Taylor's theorem to prove that

(i) $$e^{x + h} = e^x \left[1 + h + \frac{h^2}{2!} + \frac{h^3}{3!} + \cdots \right]$$

(ii) $$\frac{1}{x + h} = \frac{1}{x} \left[1 - \frac{h}{x} - \frac{h^2}{x^2} - \frac{h^3}{x^3} \cdots \right]$$

(iii) $$\log (x + h) = \log x + \frac{h}{x} - \frac{h^2}{2x^2} + \frac{h^3}{3x^3} - \cdots$$

(iv) $$\sec^{-1} (x + h) = \sec^{-1} x + \frac{h}{x\sqrt{x^2 - 1}} - \frac{h^2}{2!} \frac{2x^2 - 1}{x^2 (x^2 - 1)^{3/2}} + \cdots$$

15. (i) Expand e^x in ascending powers of $(x - 1)$.

(ii) Expand $2 + x^2 - 3x^5 + 7x^6$ in powers of $(x - 1)$.

16. (i) Expand $\log (\sec x + \tan x)$

(ii) Expand $\sec x$.

17. Assuming the possibility of expansion, prove that

(i) $$e^{\sin x} = 1 + x + \frac{1}{2}x^2 - \frac{1}{8}x^4 + \ldots$$

(ii) $$e^{x \cos x} = 1 + x + \frac{x^2}{2} - \frac{x^3}{3} + \ldots$$

(iii) $$e^x \log(1+x) = x + \frac{x^2}{2} + \frac{x^3}{3} + \frac{3x^5}{40}$$

18. Obtain by Maclaurin's theorem the first five terms in the expansion of $\log(1 + \sin x)$.

19. Obtain by Maclaurin's theorem the expansion of $\log(1 + \sin^2 x)$ upto x^4.

20. Assuming validity of expansion, show that

(a) $$\log[1 + \log(1+x)] = x - x^2 + \frac{7}{6}x^3 + \ldots$$

(b) $$\log \frac{\tan x}{x} = \frac{x^2}{3} + \frac{7}{90}x^4 + \ldots\ldots$$

(c) $$\tan^{-1}(1+x) = \frac{\pi}{4} + \frac{1}{2}x - \frac{1}{4}x^2 + \frac{1}{12}x^3 + \ldots$$

(d) $$\tan^{-1}\frac{2x}{1-x^2} = 2\left(x - \frac{x^3}{3} + \frac{x^5}{5}\ldots\right)$$

21. If $f(x)$ is continuous in $[a, b]$ and possesses finite first and second derivatives, at $c \in (a, b)$, prove that

$$f''(c) = \lim_{h \to 0} \frac{f(c+h) + f(c-h) - 2f(c)}{h^2}$$

22. (a) if $y = \tan^{-1} x$, prove that

(i) $$y_n = \frac{(-1)^{n-1}(n-1)!}{(x+1)^{n/2}} \sin\left(n \tan^{-1}\frac{1}{x}\right)$$

(ii) $$y_n = (n-1)! \sin n\left(\frac{\pi}{2} - y\right) \sin^n\left(\frac{\pi}{2} - y\right).$$

23. Find n^{th} derivatives of the following functions :

(a) $$y = \frac{x}{(x+1)^4}$$

(b) $$y = \frac{ax+b}{cx+d}$$

(c) $$y = \frac{x}{(3x-5)(1-4x^2)}$$

(d) $$y = \frac{x^4}{(x-1)(x-2)}$$

(e) $$y = \frac{1}{1 + x + x^2 + x^3}$$

(f) $$y = \frac{x^2}{(x-1)(x-2)}$$

(g) $y = \sin^{-1}\left[\dfrac{2x}{1+x^2}\right]$ (h) $y = \cos x \cos 2x \cos 3x$

(f) $y = \sin(px) + \cos(px)$ (j) $y = \dfrac{x}{(x+2)(2x+3)}$

(k) $y = \cos^{-1}\left[\dfrac{x - x^{-1}}{x + x^{-1}}\right]$ (l) $y = \tan^{-1}\left[\dfrac{\sqrt{1+x^2} - 1}{x}\right]$

(m) $y = \dfrac{x^2 + x + 1}{x^3 - 6x^2 + 11x - 6}$

24. (a) Find n^{th} derivative of $y = \tan^{-1} x$. Hence, prove that the value of $D^n(\tan^{-1} x)$ when $x = 0$ is 0, $(n-1)!$ or $-(n-1)!$ according as n is of the form $2p$, $(4p+1)$ or $(4p+3)$ respectively.

(b) Prove that, $\dfrac{d^n}{dx^n}[x^{n-1} \log x] = \dfrac{(n-1)!}{x}$.

(c) If $y = x \log(x+1)$, prove that $y_n = \dfrac{(-1)^{n-2}(n-2)!(x+n)}{(x+1)^n}$.

(d) If $y = \dfrac{ax+b}{cx+d}$, show that $y_1 y_3 = 3y_2^2$.

(e) If $x = \sin t$, $y = \sin pt$, prove that $(1 - x^2)\dfrac{d^2y}{dx^2} - x\dfrac{dy}{dx} + p^2 y = 0$.

(f) If $I_n = \dfrac{d^n}{dx^n}(x^n \log x)$, prove that :

(i) $I_n = nI_{n-1} + (n-1)!$ (ii) $\dfrac{I_n}{n!} = \dfrac{1}{n} + \dfrac{I_{n-1}}{(n-1)!}$

(iii) $I_n = n!\left[\log x + 1 + \dfrac{1}{2} + \dfrac{1}{3} + \ldots + \dfrac{1}{n}\right]$.

(g) Prove that the value of nth differential coefficient of $\dfrac{x^3}{(x^2-1)}$ for $x = 0$ is zero if n even and $-n!$ if n is odd and greater than 1.

(h) If $f(x) = \tan x$, then prove that

$$f^n(0) - {}^nC_2 - f^{n-2}(0) + {}^nC_4 f^{n-4}(0) + \ldots + \ldots = \sin\left(\dfrac{n\pi}{2}\right).$$

25. (a) State Leibnitz's theorem and find the nth derivatives of following functions :

(i) $x^2 e^x \cos x$ (ii) $x^2 e^{3x} \sin 4x$

(iii) $e^x (2x+3)^3$ (iv) $x^2 e^x$

(b) If $f(x) = \tan x$, then prove that

$$f^n(0) - {}^nC_2 - f^{n-2}(0) + {}^nC_4 f^{n-4}(0) + \ldots + \ldots = \sin\left(\frac{n\pi}{2}\right).$$

26. (a) If $y = \sin 2\theta$, $x = \sin \theta$, show that $(1 - x^2)y_{n+2} - (2n + 1)xy_{n+1} - (n^2 - 4)\, y_n = 0$

(b) If $x = \sin t$, $y = \cos wt$, show that

   (i) $(1 - x^2)\, y_1^2 = w^2 (1 - y^2)$

   (ii) $(1 - x^2)\, y_{n+2} - (2n + 1)\, xy_{n+1} - (n^2 - w^2)\, y_n = 0$

(c) Find the value of y_n at $x = 0$, if $y = e^{a \sin^{-1} x}$

(d) If $y = e^{m \cos^{-1}}$, prove that

   $(1 - x^2)\, y_{n+2} - (2n + 1)\, xy_{n+1} - (n^2 + m^2) = 0$. Hence evaluate $(y_n)_0$.

(e) If $y = \left[x + \sqrt{x^2 - 1}\right]^m$, prove that

   $(x^2 - 1)y_{n+2} + (2n + 1)xy_{n+1} + (n^2 - m^2)y_n = 0$.

(f) If $y = A \cos h(\log x^m) + B \sin h(\log x^m)$, prove that

   $x^2 y_{n+2} + (2n + 1)xy_{n+1} + (n^2 - m^2)y_n = 0$.

(g) If $x = \tan(\log y)$, prove that

   $(1 + x^2)y_{n+1} + (2nx + 1)y_n + n(n - 1)y_{n-1} = 0$

(h) If $y = \cos(m \log x)$, prove that

   $x^2 y_{n+1} + (2n + 1)xy_{n+1} + (m^2 + n^2)y_n = 0$

(i) If $y = e^{a \sin^{-1} x}$, prove that

   $(1 - x^2)y_{n+2} - (2n + 1)xy_{n+1} + (m^2 + a^2)y_n = 0$

(j) If $y = \sin^{-1} x$, prove that $(1 - x^2)y_{n+2} - (2n + 1)xy_{n+1} - n^2 y_n = 0$

(k) If $y = \sin(m \sin^{-1} x)$, prove that

   $(1 - x^2)y_{n+1} - (2n + 1)xy_{n+1} - (n^2 - m^2)y_n = 0$. Also find $(y_n)_0$.

27. Prove that :

(a) $\log \sec(x) = \dfrac{x^2}{2} + \dfrac{1}{3}\dfrac{x^4}{4} + \dfrac{2}{15}\dfrac{x^6}{6} + \ldots$

(b) $\cos x \cos h\, x = 1 - \dfrac{2^2 x^4}{4!} + \dfrac{2^4 x^8}{8!} + \ldots$

(c) $\log(1 + \sin x) = x - \dfrac{x^2}{2} + \dfrac{x^3}{6} - \dfrac{x^4}{12} + \ldots$

(d) $e^{e^x} = e\left[1 + x + x^2 + \dfrac{5}{6}x^3 + \dfrac{5}{8}x^4 + \ldots\right]$

(e) $\sqrt{1 + \sin x} = 1 + \dfrac{x}{2} - \dfrac{x^2}{8} - \dfrac{x^3}{48} + \dfrac{x^4}{348} - \cdots$

(f) $x \csc x = 1 + \dfrac{x^2}{6} - \dfrac{7}{360} x^4 + \cdots$

(g) $\dfrac{x}{e^x - 1} = 1 - \dfrac{x}{2} + \dfrac{x^2}{12} - \dfrac{x^4}{720} + \cdots$

(h) $e^x \cos x = 1 + x + \dfrac{x^2}{2} - \dfrac{x^3}{3} - \dfrac{11}{24} x^4 + \cdots$

(i) $e^x \cos x = 1 + x - \dfrac{x^3}{3} - \dfrac{x^4}{6} + \cdots$

(j) $\sin x \cosh x = x + \dfrac{x^3}{3} - \dfrac{x^5}{30} + \cdots$

(k) $\sqrt{\dfrac{1 + e^x}{2e^x}} = 1 - \dfrac{1}{4} x + \dfrac{3}{32} x^2 - \cdots$

(l) $e^{x \cos x} \cos (x \sin \alpha) = \displaystyle\sum_{n=0}^{\infty} \dfrac{x^n}{n!} \cos (n\alpha)$

(m) $(1 + x)^{1/x} = e\left[1 - \dfrac{x}{2} + \dfrac{11}{24} x^2 + \cdots \right]$

(n) $\log (1 + \sin x) = x^2 - \dfrac{5}{6} x^4 + \dfrac{32}{45} x^6 + \cdots$

(o) $\log\left[\dfrac{1 + e^{2x}}{e^x} \right] = \log 2 + \dfrac{x^2}{2} - \dfrac{x^4}{12} + \dfrac{x^6}{45} - \cdots$

(p) $\log \tan \left(\dfrac{\pi}{4} + x \right) = 2x + \dfrac{4}{3} x^3 + \dfrac{4}{3} x^5 + \cdots$

(q) $\tan^{-1} x = \dfrac{\pi}{4} + \dfrac{1}{2} (x - 1) - \dfrac{(x - 1)^2}{4} + \dfrac{(x - 1)^3}{12} + \cdots$

(r) $\sec^{-1}\left[\dfrac{1}{1 - 2x^2} \right] = 2\left[n\pi + x + \dfrac{x^3}{6} + \dfrac{3x^5}{40} + \cdots \right]$

(s) $\cos^{-1} (\tanh (\log x)) = \pi - 2\left(x - \dfrac{x^3}{3} + \dfrac{x^5}{5} - \dfrac{x^7}{7} + \cdots \right)$

(t) $e^x = 1 + \tan x + \dfrac{1}{2!} \tan^2 x - \dfrac{1}{3!} \tan^3 x - \dfrac{7}{4!} \tan^4 x + \cdots$

(u) $f\left(\dfrac{x^2}{1 + x} \right) = f(x) - \dfrac{x}{1 + x} f'(x) + \dfrac{1}{2!} \dfrac{x^2}{(1 + x)^2} f''(x) + \cdots$

(v) $\tan^{-1}\left[\dfrac{x \sin \theta}{1 - x \cos \theta}\right] = x \sin \theta + \dfrac{x^2}{2} \sin 2\theta + \dfrac{x^3}{2} \sin 3\theta + \ldots$

(w) $\dfrac{1}{2}[f(x) - f(2a - x)] = (x - a)\, f'(a) + \dfrac{(x-a)^3}{3!}\, f'''(a) + \dfrac{(x-a)^5}{5!}$

$$f^{v}(a) + \ldots$$

(x) $\tan^{-1}\left[\dfrac{\sqrt{1 + x^2} - 1}{x}\right] = \dfrac{1}{2}\left(x - \dfrac{x^3}{3} + \dfrac{x^5}{5} - \ldots\right)$

$$= \dfrac{1}{2}\left(n\pi + x - \dfrac{x^3}{3} + \dfrac{x^5}{5} - \dfrac{x^7}{7} + \ldots\right)$$

28. Expand the following functions :

    (a) $(1 + x)^x$ in a series upto a term in x^4.

    (b) $\log (1 + x + x^2 + x^3)$ upto x^8

    (c) $\sin^{-1}(3x - 4x^3)$

    (d) $\sinh^{-1}(3x + 4x^3)$

29. Using Taylor's theorem, find the expansion of the following functions in ascending powers of x :

    (i) $\tan\left[x + \dfrac{\pi}{4}\right]$ upto terms in x^4 and find approximately the value of $\tan(43°)$.

    (ii) $\log \cos\left(x + \dfrac{\pi}{4}\right)$

30. (i) Express :

    (a) $(x - 2)^4 - 3(x - 2)^3 + 4(x - 2)^2 + 5$ in powers of x

    (b) $2x^3 + 7x^2 + x - 6$ in ascending powers of $(x - 2)$

    (c) $49 + 69x + 42x^2 + 11x^3 + x^4$ in powers of $(x + 2)$

    (d) $7 + (x + 2) + 3(x + 2)^2 + (x + 2)^4 - (x + 2)^5$ in ascending powers of x.

    (e) $x^4 - 3x^3 + 2x^2 - x + 1$ in powers of $(x - 3)$

    (f) $(x + 2)^4 + 5(x + 2)^3 + 6(x + 2)^2 + 7(x + 2) + 8$ in ascending powers of $(x - 1)$.

    (g) $2x^3 + 3x^2 - 8x + 7$ in ascending powers of $(x - 2)$

  (ii) (a) If $x = (1 - y)(1 - 2y)$, then show that $y = 1 + x - 2x^3 + \ldots$

     (b) If $x^3 + 2xy^2 - y^3 + x = 1$, obtain the expression of y in ascending powers of x.

31. Evaluate the following limits :

(1) $\displaystyle\lim_{x\to\infty} \dfrac{e^{1/x} + e^{2/x} + \dots + e^{x/x}}{x}$

(2) $\displaystyle\lim_{x\to\infty} \left(x + \dfrac{1}{2}\right)\left\{\log\left(x + \dfrac{1}{2}\right) - \log x\right\}$

(3) $\displaystyle\lim_{x\to0} \dfrac{\log \tan x}{\log x}$

(4) $\displaystyle\lim_{x\to0} \left(\dfrac{a^x + b^x}{2}\right)^{1/x}$

(5) $\displaystyle\lim_{x\to1} (1 - x) \tan \dfrac{\pi x}{2}$

(6) $\displaystyle\lim_{x\to1} \dfrac{x \log x - (x - 1)}{(x - 1) \log x}$

(7) $\displaystyle\lim_{x\to0} \left(\dfrac{1}{x}\right)^{2\sin x}$

(8) $\displaystyle\lim_{x\to0} \dfrac{(1 + x)^{1/x} - e}{x}$

(9) $\displaystyle\lim_{x\to0} \dfrac{\log \tan 2x}{\log \tan x}$

(10) $\displaystyle\lim_{x\to y} \dfrac{x^y - y^x}{x^x - y^y}$

(11) $\displaystyle\lim_{x\to0} (\operatorname{cosec} x)^{\sin x}$

(12) $\displaystyle\lim_{x\to0} \dfrac{1 - x^x}{x \log x}$

(13) $\displaystyle\lim_{x\to\infty} \left[\dfrac{2^{1/x} + 3^{1/x}}{2}\right]^x$

(14) $\displaystyle\lim_{x\to0} \dfrac{e^{2x} - (1 + x)^2}{x \log (1 + x)}$

(15) $\displaystyle\lim_{x\to0} \log_x \sin 2x$

(16) $\displaystyle\lim_{x\to\pi/2} (\sin x)^{\sin^2 x}$

(17) $\displaystyle\lim_{x\to0} \dfrac{\log_{\sec x/2} \cos x}{\log_{\sec x} \cos x/2}$

(18) $\displaystyle\lim_{x\to0} \left(\dfrac{a^x + b^x}{2}\right)^{1/x}$

(19) $\displaystyle\lim_{x\to0} \left\{\dfrac{a}{x} - \cot \dfrac{x}{a}\right\}$

(20) $\displaystyle\lim_{x\to0} \left[\dfrac{x}{x - 1} - \dfrac{1}{\log x}\right]$

(21) $\displaystyle\lim_{x\to0} \dfrac{\log \tan 3x}{\log \tan 2x}$

(22) $\displaystyle\lim_{x\to0} \left[\dfrac{e^x - e^{-x} - 2 \log (1 + x)x}{x \sin x}\right]$

(23) $\displaystyle\lim_{x\to1} (1 - x^2)^{\frac{1}{\log (1 - x)}}$

(24) $\displaystyle\lim_{x\to0} \dfrac{(1 + x)^n - 1}{x}$

(25) $\displaystyle\lim_{x\to1} \left(x \sin \dfrac{1}{x}\right)^{x^2}$

(26) $\displaystyle\lim_{x\to0} \left[\dfrac{f'(x)}{f(x) - f(a)} - \dfrac{1}{x - a}\right]$

(27) $\displaystyle\lim_{x\to2} \left[\dfrac{1}{x - 2} - \dfrac{1}{\log (x - 1)}\right]$

(28) $\displaystyle\lim_{x\to0} \dfrac{(1 + x)^{1/x} - e}{x}$

(29) $\displaystyle\lim_{x\to 0} \frac{\log(1 + kx^2)}{1 - \cos x}$

(30) $\displaystyle\lim_{x\to 0} (\cos x)^{\cot x}$

(31) $\displaystyle\lim_{x\to 0} \sqrt{\frac{a + x}{a - x}} \tan^{-1}\sqrt{a^2 - x^2}$

(32) $\displaystyle\lim_{x\to\infty} \left\{ x - x^2 \log\left[1 + \frac{1}{x}\right] \right\}$

(33) $\displaystyle\lim_{x\to 0} (\cos x)^{\cos^2 x}$

(34) $\displaystyle\lim_{x\to 0} \left[\frac{2^x + 3^x}{2}\right]^{1/x}$

(35) $\displaystyle\lim_{x\to 0} \left[\frac{\pi}{4x} - \frac{\pi}{2x(e^{px} + 1)}\right]$

(36) $\displaystyle\lim_{x\to\infty} \left(\cos\left(\frac{\theta}{n}\right)\right)^{n^2}$

(37) $\displaystyle\lim_{x\to 0} \frac{A}{x^2}\left[\frac{\sin Kx}{\sin Lx} - \frac{K}{L}\right]$

(38) $\displaystyle\lim_{x\to 0} \left[\sin^2\left(\frac{\pi}{2 - ax}\right)\right]^{\sec^2\left(\frac{\pi}{2 - ax}\right)}$

(39) $\displaystyle\lim_{x\to 0} \left[\frac{a^x + b^x + c^x + d^x}{4}\right]^{1/x}$

(40) $\displaystyle\lim_{x\to 0} \left(\frac{1}{x}\right)^{2\sin x}$

(41) $\displaystyle\lim_{x\to 0} \frac{1}{2}\left[\sqrt{\frac{a}{x}} + \sqrt{\frac{x}{a}}\right]^{\frac{1}{(x - a)}}$

(42) $\displaystyle\lim_{x\to 0} \frac{x \sin(\sin x) - \sin^2 x}{x^6}$

32. (a) Find the value of a, b and c, if :

(i) $\displaystyle\lim_{x\to 0} \frac{x(a + b \cos x) - c \sin x}{x^5} = 1$

(ii) $\displaystyle\lim_{x\to 0} \frac{a \tanh x + b \sin x + cx}{x^5} = \frac{7}{60}$

(iii) $\displaystyle\lim_{x\to 0} \frac{\theta(a + b \cos \theta)}{\theta^5} = 1$

(iv) $\displaystyle\lim_{x\to 0} \frac{ae^x - b \cos x + ce^{-x}}{x \sin x} = 2$

Answers

[A] (1) False   (2) False   (3) False   (4) False   (5) False   (6) True

[B] (1) - (c)   (2) - (d)   (3) - (c)   (4) - (d)   (5) - (c)   (6) - (b)   (7) - (c)
(8) - (b)   (9) - (c)   (10) - (c)

[D]

1. (a) $\dfrac{3}{2}$　　(b) 1　　(c) $\dfrac{1}{2}$　　(d) $\dfrac{\pi^2}{2e}$

(e) $\dfrac{\log a}{\log b}$　　(f) 1　　(g) 2　　(h) 4

(i) 1　　(j) 1　　(k) $\dfrac{1}{5}$　　(l) 0

(m) $\dfrac{1}{2}$　　(n) $\dfrac{2}{3}$　　(o) $-\dfrac{1}{2}$　　(p) 0

(q) 0　　(r) $\dfrac{2a}{\pi}$　　(s) $e^{-1/2}$　　(t) 1

(u) $\log a$　　(v) 1　　(w) e　　(x) $\dfrac{1}{e}$

(y) $\dfrac{1}{e}$

2. $a = -\dfrac{5}{2}$; $b = -\dfrac{3}{2}$ 　　　　3. limit $= -1$, $a = -2$.

4. (i) $-\dfrac{1}{2}$, (ii) $\dfrac{1}{2}$, (iii) 0, (iv) 0, (v) $-\dfrac{1}{3}$, (vi) $-\dfrac{1}{2}$, (vii) ab,

(viii) $e^{-\frac{a^2 b}{2}}$, (ix) e^8, (x) 1.

5. (i) $(-1)^n\, n! \left[\dfrac{16}{(x-2)^{n+1}} - \dfrac{1}{(x-1)^{n+1}} \right]$, $n > 2$

(ii) $\dfrac{(-1)^n\, n!\, 9.2^{n-1}}{(2x+3)^{n+1}} - \dfrac{(-1)^n\, n!\, 4}{(x+2)^{n+1}}$

(iii) $\dfrac{(-1)^n\, n!}{4} \left[\dfrac{3}{(x+2)^{n+1}} + \dfrac{1}{(x+2)^{n+1}} \right]$

(iv) $\dfrac{1}{4} \left[\sin\left(x + \dfrac{n\pi}{2}\right) + 3^n \sin\left(3x + \dfrac{n\pi}{2}\right) \right]$

(v) $\dfrac{1}{4} \left[2^n \cos\left(2x + \dfrac{n\pi}{2}\right) + 4^n \cos\left(4x + \dfrac{n\pi}{2}\right) + 6^n \cos\left(6x + \dfrac{n\pi}{2}\right) \right]$

(vi) $e^x \left[x^n + \dfrac{n^2}{1!} x^{n-1} + \dfrac{n^2\,(n-1)^2}{2!} x^{n-2} + \dots \right]$

(vii) $e^x \left[\log x + n_{c_1} x^{-1} - n_{c_2} x^{-2} + \dots\dots + (-1)^{n-1} (n-1)!\, x^{-n} \right]$

(viii) $2^{(n-2)/2}$

e^x

$$\left[2x^2 \cos\left(x + \frac{n\pi}{4}\right) + 2^{3/2}\, nx \cos\left(x + (n-1)\frac{\pi}{4}\right) + n(n-1)\cos\left(x + (n-2)\frac{\pi}{4}\right)\right]$$

(ix) $\dfrac{1}{2}\, 2^n \cos\left(2x + \dfrac{n\pi}{2}\right) + \dfrac{1}{8}\, 4^n \cos\left(4x + \dfrac{n\pi}{2}\right)$

15. (i) $\left[1 + (x-1) + \dfrac{(x-1)^2}{2!} + \dfrac{(x-1)^3}{3!} + \ldots \right]$

(ii) $7 + 29\,(x-1) + 76\,(x-1)^2 + 110\,(x-1)^3 + 90\,(x-1)^4$
$$+\; 39\,(x-1)^5 + 7(x-1)^6$$

16. (i) $x + \dfrac{x^3}{6} + \dfrac{x^5}{24} + \ldots$ (ii) $1 + \dfrac{x^2}{2!} + \dfrac{5x^4}{4!} + \ldots$

18. $x - \dfrac{x^2}{2!} + \dfrac{x^3}{3!} - \dfrac{2x^4}{4!} + \dfrac{5x^5}{5!} - \ldots$

19. $x^2 - \dfrac{5}{6}x^4 + \ldots$

✍ ✍ ✍

Chapter 3...
Differential Equations of First Order and First Degree

Johann Bernoulli

Johann Bernoulli (27 July 1667 – 1 January 1748) was a Swiss mathematician and was one of the many prominent mathematicians in the Bernoulli family. He is known for his contributions to infinitesimal calculus and educating Leonhard Euler in the pupil's youth. Throughout Johann Bernoulli's education at Basel University the Bernoulli brothers worked together spending much of their time studying the newly discovered infinitesimal calculus.

The Bernoulli brothers often worked on the same problems, but not without friction. Their most bitter dispute concerned the brachistochrone curve problem, or the equation for the path followed by a particle from one point to another in the shortest amount of time, if the particle is acted upon by gravity alone. Johann presented the problem in 1696, offering a reward for its solution. Entering the challenge, Johann proposed the cycloid, the path of a point on a moving wheel, also pointing out the relation this curve bears to the path taken by a ray of light passing through layers of varied density. Jacob proposed the same solution, but Johann's derivation of the solution was incorrect, and he presented his brother Jacob's derivation as his own.

3.1 Differential Equations

3.1.1 Introduction

To solve physical or real life problems mathematically, we should formulate the problem in mathematical model. If the physical problems is concerning with the change in the physical qualities, then we formulate the mathematical model in terms of equations involving unknown function and its derivatives. Such equations are called as differential equations.

If $y = f(x)$ be a given function, then its derivative $\dfrac{dy}{dx}$ can be interpreted as the rate of change of y w.r.t. x. In any natural process, the variables involved and their rates of changes are connected with one another by means of the basic scientific principles that governs the process. When this connection is expressed in Mathematical symbols, the result is often called a **differential equation**.

Differential equations are of fundamental importance in pure and applied sciences and engineering such as Algebra, Geometry, Mechanical system, electric circuits, structures, heat flows, diffusion of solvents or gases, rate of chemical reactions and many other areas including nuclear reactions, rate of decay etc.

3.1.2 Introduction to Function of Two, Three Variables, Homogeneous Functions, Partial Derivatives

(1) Function of One Variable :

Definition : A rule $f : \mathbb{R} \to \mathbb{R}$, which associates every real number x with a unique real number $y = f(x)$, such a rule f is called a **function of one variable x**.

Here the domain of f consists of all points on x-axis and the range of f is the set of all points y such that $y = f(x)$, x is called the independent variable and y is called dependent variable.

(2) Function of Two Variables x and y :

Definition : Let A be the set of all points in the co-ordinate plane XOY and B be the set of all real numbers. Thus $A = \mathbb{R} \times \mathbb{R}$, $B = \mathbb{R}$.

Consider a mapping $f : \mathbb{R} \times \mathbb{R} \to \mathbb{R}$, then a rule f which associates every point (x, y) in XOY plane with a unique real number z measured parallel to z-axis is called a function of two variables x and y and it is written as $z = f(x, y)$.

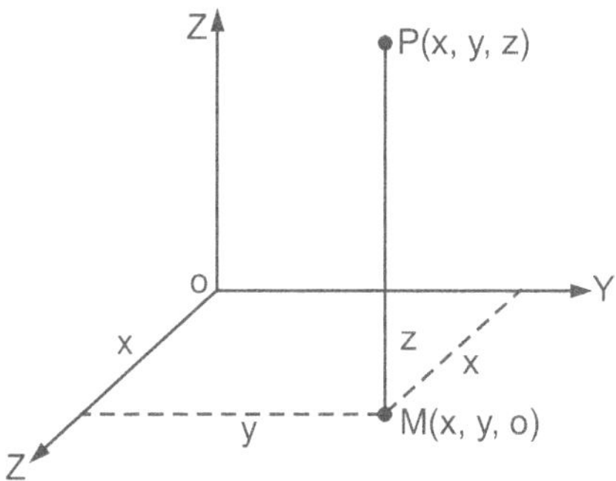

Fig. 3.1

We write $f : (x, y) \longmapsto z = f(x, y)$, $f(x, y)$ is called the value of a function f at a point (x, y) whereas x and y are called independent variables.

(3) Domain and Range of Functions of Two Variables x and y :

Definition : Domain of a function of two variables is a subset of $\mathbb{R} \times \mathbb{R}$.

$$\therefore \quad \text{Domain } (f) = \{(x, y) : (x, y) \in A \in \mathbb{R} \times \mathbb{R}\}$$

The set of all values $f(x, y)$ is called the range of f i.e.

$$\text{Range } f = R(f) = \{f(x, y) \mid (x, y) \in A\}$$

(4) Function of Three Variables x, y and z :

Definition : Let A be the set of all points in space and B be the set of all real numbers. Then the rule f, which associates to every point (x, y, z) in space with a unique real number u is called a function of three variables x, y and z and it is written as $u = f(x, y, z)$.

(5) Homogeneous Function of Two Variables :

Definition : A function f where, $u = f(x, y)$ is said to be homogeneous function of degree n, iff u is expressible as $x^n \phi \left(\dfrac{y}{x} \right)$ i.e.

$$u = x^n \phi \left(\frac{y}{x} \right), \text{ n is a real number.}$$

By another approach, we can define homogeneous function of two variables as –

Definition : A function of two variables x and y is said to be homogeneous function of degree n, iff

$$f(tx, ty) = t^n f(x, y) \ \forall \ t > 0,$$

such that (x, y) and (tx, ty) are in domain of f.

For example, (i) The function

$$f(x, y) = 5x^4 - 7x^3y + 3xy^3$$

is a homogeneous function of degree n = 4, since

$$f(x, y) = x^4 \left[5 - 7 \left(\frac{y}{x} \right) + 3 \left(\frac{y}{x} \right)^2 \right]$$

$$= x^4 \, \phi \left(\frac{y}{x} \right)$$

$$\therefore \qquad f(x, y) = x^4 \, \phi \left(\frac{y}{x} \right)$$

By another approach,

$$f(tx, ty) = 5t^4 x^4 - 7t^4 x^3y + 3t^4 xy^3$$

$$= t^4 (5x^4 - 7x^3y + 3xy^3)$$

$$= t^4 f(x, y)$$

$$\therefore \qquad f(tx, ty) = t^4 f(x, y), \ \forall \ (t > 0)$$

$\therefore$ f(x, y) is homogeneous function of degree n = 4.

(ii) The function, $\qquad u = f(x, y) = \dfrac{x^{1/5} + y^{1/5}}{x^{1/3} + y^{1/3}}$

is a homogeneous function of degree $n = -\dfrac{2}{15}$, since

$$u = \frac{x^{1/5} + y^{1/5}}{x^{1/3} + y^{1/3}}$$

$$= \frac{\left[1 + \left(\frac{y}{x} \right)^{1/5} \right]}{\left[1 + \left(\frac{y}{x} \right)^{1/3} \right]}$$

$$= x^{1/5 - 1/3} \, \frac{x^{1/5} \left[1 + \left(\frac{y}{x} \right)^{1/5} \right]}{x^{1/3} \left[1 + \left(\frac{y}{x} \right)^{1/3} \right]}$$

$$= x^{-2/15} \, \phi \left(\frac{y}{x} \right)$$

$$\therefore \qquad u = x^{-2/15} \, \phi \left(\frac{y}{x} \right)$$

By another approach,

$$f(tx, ty) = \frac{(tx)^{1/5} + (ty)^{1/5}}{(tx)^{1/3}\,(ty)^{1/3}}$$

$$= \frac{t^{1/5}}{t^{1/3}}\left[\frac{x^{1/5} + y^{1/5}}{x^{1/3} + y^{1/3}}\right]$$

$$= t^{-2/15}\, f(x, y)$$

$$\therefore \qquad f(tx, ty) = t^{-2/15}\, f(x, y)$$

$\therefore$ $u = f(x, y)$ is homogeneous function of degree $n = -\dfrac{2}{15}$

(6) Homogeneous Function of Three Variables :

Definition : A function f where $u = f(x, y, z)$ is said to be a homogeneous function of degree n, iff u is expressible as $x^n\,\phi\left(\dfrac{y}{x}, \dfrac{z}{x}\right)$ i.e.

$$u = x^n\,\phi\left(\frac{y}{x}, \frac{z}{x}\right), \text{ n is a real number}$$

By another approach we can define homogeneous function of three variables as –

Definition : A function f of three variables x, y, z is said to homogeneous function of degree n, iff

$$f(tx, ty, tz) = t^n\, f(x, y, z), n \in \mathbb{R}, \forall\, t > 0$$

such that (x, y, z) and (tx, ty, tz) are in domain of f.

For example, (i) The function

$$f(x, y, z) = 3x^3 + y^2z + z^3$$

is a homogeneous function of degree $n = 3$, since

$$f(x, y, z) = x^3\left[1 + \left(\frac{y}{x}\right)^2\left(\frac{z}{x}\right) + \left(\frac{z}{x}\right)^3\right]$$

$$= x^3\,\phi\left(\frac{y}{x}, \frac{z}{x}\right)$$

By another approach

$$f(tx, ty, tz) = 3t^3x^3 + t^2y^2 \cdot tz + t^3z^3$$

$$= t^3\,(3x^3 + y^2z + z^3)$$

$$= t^3\, f(x, y, z)$$

$$\therefore \qquad f(tx, ty, tz) = t^3\, f(x, y, z), \forall\, t > 0.$$

$\therefore$ $f(x, y, z)$ is homogeneous function of degree $n = 3$.

(ii) The function $\quad u = f(x, y, z) = \dfrac{x + y + z}{\sqrt{x} + \sqrt{y} + \sqrt{z}}$

is homogeneous function of degree $n = \dfrac{1}{2}$, since

$$u = \frac{x + y + z}{\sqrt{x} + \sqrt{y} + \sqrt{z}}$$

$$= \frac{x\left[1 + \dfrac{y}{x} + \dfrac{z}{x}\right]}{\sqrt{x}\left[1 + \sqrt{\dfrac{y}{x}} + \sqrt{\dfrac{z}{x}}\right]}$$

$$= x^{1/2}\, \phi\left(\frac{y}{x}, \frac{z}{x}\right)$$

$$\therefore \qquad u = x^{1/2}\, \phi\left(\frac{y}{x}, \frac{z}{x}\right)$$

By another approach, we have

$$f(x, y, z) = \frac{x + y + z}{\sqrt{x} + \sqrt{y} + \sqrt{z}}$$

$$\therefore \qquad f(tx, ty, tz) = \frac{x + y + z}{\sqrt{x} + \sqrt{y} + \sqrt{z}} = \frac{tx + ty + tz}{\sqrt{tx} + \sqrt{ty} + \sqrt{tz}}$$

$$= t^{1/2}\, f(x, y, z)$$

$$\therefore \qquad f(tx, ty, tz) = t^{1/2}\, f(x, y, z),\ t > 0$$

$\therefore\ \ u = f(x, y, z)$ is homogeneous function of degree $n = \dfrac{1}{2}$.

(7) Partial Derivatives :

Definition : The partial derivative of $f(x, y)$ w.r.t. x is the ordinary derivative of $f(x, y)$ when y is regarded as constant. It is written as $\dfrac{\partial f}{\partial x}$ or f_x

Thus, $\qquad \dfrac{\partial f}{\partial x} = \lim_{\delta x \to 0} \dfrac{f(x + \delta x, y)}{\delta x}$, if it exist.

Again the partial derive $\dfrac{\partial f}{\partial y}$ or f_y of $f(x, y)$ w.r.t. y is ordinary derivative of $f(x, y)$, when x is regarded as constant and we write

$$\frac{\partial f}{\partial y} = \lim_{\delta y \to 0} \frac{f(x, y + \delta y) - f(x, y)}{\delta y}\ ,\ \text{if it exist.}$$

The partial derivatives $\dfrac{\partial f}{\partial x}$ and $\dfrac{\partial f}{\partial y}$ are called the **first order partial derivatives** of $f(x, y)$

Remark : $\left(\dfrac{\partial f}{\partial x}\right)_{(x,\ y)}$ and $\left(\dfrac{\partial f}{\partial y}\right)_{(x,\ y)}$ i.e. $f_x(x,\ y)$ and $f_y(x,\ y)$ are new functions of x and y.

Higher Order Partial Derives :

If $f(x,\ y)$ is a function of two variables x and y, then the first partial derivatives of $f(x,\ y)$ w.r.t. x and w.r.t. y are $\dfrac{\partial f}{\partial x}$, $\dfrac{\partial f}{\partial y}$ respectively. These are also written as f_x, f_y respectively.

$\dfrac{\partial f}{\partial x}$ and $\dfrac{\partial f}{\partial y}$ are themselves functions of x and y which may have partial derivatives w.r.t x and w.r.t. y and these are

$$\frac{\partial}{\partial x}\left(\frac{\partial f}{\partial x}\right) = \frac{\partial^2 f}{\partial x^2} = f_{xx}$$

$$\frac{\partial}{\partial x}\left(\frac{\partial f}{\partial y}\right) = \frac{\partial^2 f}{\partial x\ \partial y} = f_{xy}$$

$$\frac{\partial}{\partial y}\left(\frac{\partial f}{\partial x}\right) = \frac{\partial^2 f}{\partial y\ \partial x} = f_{yx}$$

$$\frac{\partial}{\partial y}\left(\frac{\partial f}{\partial y}\right) = \frac{\partial^2 f}{\partial y^2} = f_{yy}$$

Thus, there are four second order partial derivatives of $f(x,\ y)$. The second order partial derivatives $\dfrac{\partial^2 f}{\partial x\ \partial y}$ and $\dfrac{\partial^2 f}{\partial y\ \partial x}$ are called mixed second order partial derivatives. For the purpose of this text, we assume that $\dfrac{\partial^2 f}{\partial x\ \partial y} = \dfrac{\partial^2 f}{\partial y\ \partial x}$. The student will verify and convince himself that the mixed second order partial derivatives are equal.

Definition : If $f(x,\ y,\ z)$ is a function of three variables, and two of these variables are hold fixed, then $f(x,\ y,\ z)$ may be regarded as dependent only on remaining variables i.e. $f(x,\ y,\ z)$ is regarded as only function of one variable and it may possess derivative w.r.t this variable, such derivative if it exists, is called as **partial derivative** of $f(x,\ y,\ z)$ w.r.t. this variable.

Remark : The process of finding a derivative of a function of two or more variables w.r.t one of these variables, keeping other variables constant is called partial differentiation. It is essentially an ordinary derivative w.r.t. one of the variables concerned, keeping other variables constant.

Illustration 3.1 : Find the first order partial derivatives of the following functions :

(i) $f(x, y) = x^2y + \sin(x + y)$

(ii) $f(x, y) = \log(x + \cos y)$

(iii) $f(x, y) = \dfrac{2xy}{x^2 + y^2}$, $(x, y) \neq (0, 0)$

(iv) $f(x, y) = x^y + y^x$

Solution : (i) We have,

$$f(x, y) = x^2y + \sin(x + y) \qquad \ldots \text{(i)}$$

Differentiating equation (i), w.r.t. x. partially treating y constant, we get

$$\frac{\partial f}{\partial x} = 2xy + \cos(x + y)$$

Differentiating equation (i), w.r.t. y partially, treating x constant, we get

$$\frac{\partial f}{\partial y} = x^2 + \cos(x + y)$$

Thus, $\qquad \dfrac{\partial f}{\partial x} = 2xy + \cos(x + y)$ and $\dfrac{\partial f}{\partial y} = x^2 + \cos(x + y)$

(ii) We have,

$$f(x, y) = \log(x + \cos y) \qquad \ldots \text{(i)}$$

Differentiating equation (i), w.r.t. x partially treating y constant, we get

$$\frac{\partial f}{\partial x} = \frac{1}{x + \cos y} \frac{\partial}{\partial x}(x + \cos y)$$

$$= \frac{1}{x + \cos y}(1)$$

$$= \frac{1}{x + \cos y}$$

$$\therefore \quad \frac{\partial f}{\partial x} = \frac{1}{x + \cos y}$$

Differentiating equation (i) w.r.t. y partially, keeping x constant, we get

$$\frac{\partial f}{\partial y} = \frac{1}{x + \cos y} \frac{\partial}{\partial y} (x + \cos y)$$

$$= \frac{1}{x + \cos y} (-\sin y) = \frac{-\sin y}{x + \cos y}$$

$$\therefore \qquad \frac{\partial f}{\partial y} = \frac{-\sin y}{x + \cos y}$$

(iii) We have,

$$f(x, y) = \frac{2xy}{x^2 + y^2}, \quad (x, y) \neq (0, 0) \qquad \qquad \dots \text{(i)}$$

Differentiating equation (i), w.r.t. x partially, keeping y constant, we get

$$\frac{\partial f}{\partial x} = \frac{(x^2 + y^2) \dfrac{\partial}{\partial x} (2xy) - (2xy) \dfrac{\partial}{\partial x} (x^2 + y^2)}{(x^2 + y^2)^2}$$

$$= \frac{(x^2 + y^2)\, (2y) - (2xy)\, (2x)}{(x^2 + y^2)^2}$$

$$= \frac{2y\, (y^2 - x^2)}{(x^2 + y^2)^2}$$

$$\therefore \qquad \frac{\partial f}{\partial x} = \frac{2y\, (y^2 - x^2)}{(x^2 + y^2)^2}$$

Differentiating equation (i) w.r.t. y keeping x constant, we get

$$\frac{\partial f}{\partial y} = \frac{(x^2 + y^2) \dfrac{\partial}{\partial y} (2xy) - (2xy) \dfrac{\partial}{\partial y} (x^2 + y^2)}{(x^2 + y^2)^2}$$

$$= \frac{(x^2 + y^2)\, (2x) - (2xy)\, (2y)}{(x^2 + y^2)^2} = \frac{2x\, (x^2 - y^2)}{(x^2 + y^2)^2}$$

$$\therefore \qquad \frac{\partial f}{\partial y} = \frac{2x\, (x^2 - y^2)}{(x^2 + y^2)^2}$$

(8) Total Derivative :

If u = f(x, y), where x = ϕ(t) and y = ϕ(t), then u can be expressed as a function of the single variable t, if we substitute for x and y in u = f(x, y) and then the derivative of u w.r.t. t is ordinary differential coefficient $\dfrac{du}{dt}$.

This $\dfrac{du}{dt}$ is called the **Total Derivative**, to distinguish it from the partial derivatives $\dfrac{\partial u}{\partial x}$ and $\dfrac{\partial u}{\partial y}$. We shall now state the relation between $\dfrac{du}{dt}$ and the partial derivatives $\dfrac{\partial u}{\partial x}, \dfrac{\partial u}{\partial y}$.

$\boxed{\textbf{Theorem 1}}$ **(Without Proof) :**

Let u be a composite function of t given by relation u = f(x, y), where, x = ϕ(t), y = ψ(t). If u possesses continuous partial derivatives w.r.t. x and y and x, y possesses derivatives w.r.t. t, then

$$\frac{du}{dt} = \frac{\partial u}{\partial x}\frac{dx}{dt} + \frac{\partial u}{\partial y}\frac{dy}{dt} \qquad \ldots \text{(i)}$$

An important result :   ∎

If we put t = x in equation (i), we get

$$\frac{du}{dx} = \frac{\partial u}{\partial x}\cdot\frac{dx}{dx} + \frac{\partial u}{\partial y}\frac{dy}{dx}$$

$$\frac{du}{dx} = \frac{\partial u}{\partial x}dx + \frac{\partial u}{\partial y}dy$$

i.e.
$$du = \frac{\partial u}{\partial x}dx + \frac{\partial u}{\partial y}dy$$

Similarly, if t = y in equation (i), then we can also have

$$du = \frac{\partial u}{\partial x}dx + \frac{\partial u}{\partial y}dy.$$

Some times the total derivative is also, defined as;

Let u = f(x, y) be differentiable function of x and y. Then the total differential or simply differential of u = f(x, y) is defined as;

$$du = \frac{\partial u}{\partial x}dx + \frac{\partial u}{\partial y}dy$$

where, dx and dy are differentials of x and y.

Illustrative Examples

Example 3.1 : *Which of the following functions are homogeneous functions? If they are homogeneous functions, then determine their degree ?*

(i) $f(x, y) = \dfrac{x^3}{y^3} \sin\left(\dfrac{y}{x}\right)$

(ii) $f(x, y) = \sqrt{x^2 + y^2} \, \sin^{-1}\left(\dfrac{y}{x}\right)$

(iii) $u = \sin^{-1}\left(\dfrac{x + y}{\sqrt{x} + \sqrt{y}}\right)$

(iv) $u = \dfrac{x^2 y^2 z^2}{x^2 + y^2 + z^2}$

(v) $u = \cos^{-1}\left(\dfrac{xy + yz}{x^2 + y^2 + z^2}\right)$

Solution : (i) We have, $f(x, y) = \dfrac{x^3}{y^3} \sin\left(\dfrac{y}{x}\right)$

$$\therefore \qquad f(tx, ty) = \dfrac{t^3 x^3}{t^3 y^3} \sin\left(\dfrac{ty}{tx}\right)$$

$$= \dfrac{x^3}{y^3} \sin\left(\dfrac{y}{x}\right)$$

$$= f(x, y)$$

$$= x^0 \, f(x, y)$$

$$\therefore \qquad f(tx, ty) = x^0 \, f(x, y)$$

$\therefore$ $f(x, y)$ is homogeneous function of degree $n = 0$.

(ii)   We have, $f(x, y) = \sqrt{x^2 + y^2} \, \sin^{-1}\left(\dfrac{y}{x}\right)$

$$\therefore \qquad f(tx, ty) = \sqrt{t^2 x^2 + t^2 y^2} \, \sin^{-1}\left(\dfrac{ty}{tx}\right)$$

$$= t^{\frac{1}{2}} \sqrt{x^2 + y^2} \, \sin^{-1}\left(\dfrac{y}{x}\right)$$

$$= t^{\frac{1}{2}} \, f(x, y)$$

$$\therefore \qquad f(tx, ty) = t^{\frac{1}{2}} \, f(x, y)$$

$\therefore$ $f(x, y)$ is homogeneous function of degree $n = 1$.

(iii)   Let, $u = f(x, y) = \sin^{-1}\left(\dfrac{x + y}{\sqrt{x} + \sqrt{y}}\right)$

$$\therefore \qquad f(tx, ty) = \sin^{-1}\left(\dfrac{tx + ty}{\sqrt{tx} + \sqrt{ty}}\right)$$

$$= \sin^{-1}\left(\frac{t(x+y)}{\sqrt{t}(\sqrt{x}+\sqrt{y})}\right)$$

$$= \sin^{-1}(t^{1/2} f(x, y))$$

$$\neq t^n f(x, y), \text{ for any } n$$

$\therefore$ $f(x, y)$ is not homogeneous function.

(iv) Let, $u = f(x, y, z) = \dfrac{x^2 y^2 z^2}{x^2 + y^2 + z^2}$

$\therefore$ $f(tx, ty, tz) = \dfrac{(tx)^2 (ty)^2 (tz)^2}{(tx)^2 + (ty)^2 + (tz)^2}$

$$= \frac{t^6 (x^2 y^2 z^2)}{t^2 (x^2 + y^2 + z^2)}$$

$$= t^4 f(x, y, z)$$

$\therefore$ $f(tx, ty, tz) = t^4 (f(x, y, z))$

$\therefore$ $y = f(x, y, z)$ is homogeneous function of degree $n = 4$.

(v) Let, $u = \cos^{-1}\left(\dfrac{xy + yz}{x^2 + y^2 + z^2}\right) = f(x, y, z)$, say

$\therefore$ $f(tx, ty, tz) = \cos^{-1}\left[\dfrac{t^2 (xy + yz)}{t^2 (x^2 + y^2 + z^2)}\right]$

$$= \cos^{-1}\left(\frac{xy + yz}{x^2 + y^2 + z^2}\right)$$

$$= f(x, y, z)$$

$$= t^0 f(x, y, z)$$

$\therefore$ $f(tx, ty, tz) = t^0 f(x, y, z)$

$\therefore$ $u = f(x, y, z)$ is homogeneous function of degree $n = 0$.

Example 3.2 : *Find* $\dfrac{\partial u}{\partial x}$ *and* $\dfrac{\partial u}{\partial y}$ *of the following functions :*

(i) $u = \log(x^2 + y^2 + z^2)$

(ii) $u = e^x \sin(xy)$

(iii) $u = y^3 x - 3xy^2 + z^3$

(iv) $u = \sin^{-1}\left(\dfrac{x}{y}\right)$

(v) $u = e^{-x} \cos y$

Solution : (i) We have, $u = \log(x^2 + y^2 + z^2)$ 　　　　　 ... (i)

Differentiating equation (i) w.r.t. x, treating y and z constant, we get

$$\frac{\partial u}{\partial x} = \frac{1}{x^2 + y^2 + z^2} \cdot \frac{\partial}{\partial x}(x^2 + y^2 + z^2)$$

$$= \frac{2x}{x^2 + y^2 + z^2}$$

$$\therefore \qquad \frac{\partial u}{\partial x} = \frac{2x}{x^2 + y^2 + z^2}$$

Differentiating equation (i) w.r.t. y treating x and z constant, we get

$$\frac{\partial u}{\partial y} = \frac{1}{x^2 + y^2 + z^2} \frac{\partial}{\partial y}(x^2 + y^2 + z^2)$$

$$= \frac{2y}{x^2 + y^2 + z^2}$$

$$\therefore \qquad \frac{\partial u}{\partial y} = \frac{2y}{x^2 + y^2 + z^2}$$

Differentiating equation (i) w.r.t. z treating x and y constant, we get

$$\frac{\partial u}{\partial z} = \frac{1}{x^2 + y^2 + z^2} \frac{\partial}{\partial z}(x^2 + y^2 + z^2)$$

$$= \frac{2z}{x^2 + y^2 + z^2}$$

$$\therefore \qquad \frac{\partial u}{\partial z} = \frac{2z}{x^2 + y^2 + z^2}$$

(ii)　We have, 　　　　　 $u = e^x \sin(xy)$

$$\therefore \qquad \frac{\partial u}{\partial x} = e^x \cos(xy)y + e^x \sin(xy)$$

$$= e^x [y \cos(xy) + \sin(xy)]$$

$$\therefore \qquad \frac{\partial u}{\partial x} = e^x [y \cos(xy) + \sin(xy)]$$

and 　　　　　 $$\frac{\partial u}{\partial y} = e^x \cos(xy) \cdot (x) + 0 \cdot \sin(xy)$$

$$= x\, e^x \cos(xy)$$

$$\therefore \qquad \frac{\partial u}{\partial y} = x\, e^x \cos(xy)$$

(iii) We have, $u = y^3x - 3xy^2 + z^3$

$\therefore \quad \dfrac{\partial u}{\partial x} = y^3 - 3y^2$

$\dfrac{\partial u}{\partial y} = 3y^2x - 6xy$

$\dfrac{\partial u}{\partial z} = 3z^2$

(iv) We have, $u = \sin^{-1}\left(\dfrac{x}{y}\right)$

$\therefore \quad \dfrac{\partial u}{\partial x} = \dfrac{1}{\sqrt{1 - \dfrac{x^2}{y^2}}} \cdot \dfrac{\partial}{\partial x}\left(\dfrac{x}{y}\right)$

$\qquad = \dfrac{y}{\sqrt{y^2 - x^2}} \cdot \left(\dfrac{1}{y}\right)$

$\qquad = \dfrac{1}{\sqrt{y^2 - x^2}}$

$\therefore \quad \dfrac{\partial u}{\partial x} = \dfrac{1}{\sqrt{y^2 - x^2}}$

and $\quad \dfrac{\partial u}{\partial y} = \dfrac{1}{\sqrt{1 - \dfrac{x^2}{y^2}}} \cdot \dfrac{\partial}{\partial y}\left(\dfrac{x}{y}\right)$

$\qquad = \dfrac{y}{\sqrt{y^2 - x^2}} \cdot \left(\dfrac{-x}{y^2}\right)$

$\qquad = -\dfrac{x}{y\sqrt{y^2 - x^2}}$

$\therefore \quad \dfrac{\partial u}{\partial y} = -\dfrac{x}{y\sqrt{y^2 - x^2}}$

(v) We have, $u = e^{-x}\cos y$

$\therefore \quad \dfrac{\partial u}{\partial x} = -e^{-x}\cos y$

and $\quad \dfrac{\partial u}{\partial y} = -e^{-x}\sin y$

Example 3.3 : *Verify the formula* $\dfrac{\partial^2 u}{\partial x\, \partial y} = \dfrac{\partial^2 u}{\partial y\, \partial x}$ *in each of the following cases :*

(i) $u = \sin^{-1}\left(\dfrac{y}{x}\right)$

(ii) $u = \log\left(\dfrac{x^2 + y^2}{xy}\right)$

Solution : (i) We have,

$$u = \sin^{-1}\left(\frac{y}{x}\right) \qquad \ldots \text{(i)}$$

$$\therefore \qquad \frac{\partial u}{\partial x} = \frac{1}{\sqrt{1 - \dfrac{y^2}{x^2}}}\; \frac{\partial}{\partial x}\left(\frac{y}{x}\right)$$

$$= \frac{x}{\sqrt{x^2 - y^2}}\left(\frac{-y}{x^2}\right)$$

$$= -\frac{y}{x\sqrt{x^2 - y^2}}$$

$$\therefore \qquad \frac{\partial u}{\partial x} = -\frac{y}{x\sqrt{x^2 - y^2}} \qquad \ldots \text{(ii)}$$

$$\text{and} \qquad \frac{\partial u}{\partial y} = \frac{1}{\sqrt{1 - \dfrac{y^2}{x^2}}} \cdot \frac{\partial}{\partial y}\left(\frac{y}{x}\right)$$

$$= \frac{x}{\sqrt{x^2 - y^2}} \cdot \frac{1}{x}$$

$$= \frac{1}{\sqrt{x^2 - y^2}}$$

$$\therefore \qquad \frac{\partial u}{\partial y} = \frac{1}{\sqrt{x^2 - y^2}} \qquad \ldots \text{(iii)}$$

Now differentiating equation (ii) w.r.t. y, partially, we get

$$\frac{\partial}{\partial y}\left(\frac{\partial u}{\partial x}\right) = -\left[\frac{x\sqrt{x^2 - y^2}\,(1) + y\left(\dfrac{xy}{\sqrt{x^2 - y^2}}\right)}{x^2\,(x^2 - y^2)}\right]$$

$$= -\left[\frac{x(x^2 - y^2) + xy^2}{x^2(x^2 - y^2)^{3/2}}\right]$$

$$= -\left[\frac{x^3 - xy^2 + xy^2}{x^2(x^2 - y^2)^{3/2}}\right]$$

$$= -\left[\frac{x^3}{x^2(x^2 - y^2)^{3/2}}\right]$$

$$= -\left[\frac{x}{(x^2 - y^2)^{3/2}}\right] = -\frac{x}{(x^2 - y^2)^{3/2}}$$

$$\therefore \quad \frac{\partial}{\partial y}\left(\frac{\partial u}{\partial x}\right) = -\frac{x}{(x^2 - y^2)^{3/2}}$$

$$\therefore \quad \frac{\partial^2 u}{\partial y\, \partial x} = -\frac{x}{(x^2 - y^2)^{3/2}} \qquad \ldots \text{(iv)}$$

Now differentiating equation (iii) w.r.t. x partially, we get

$$\frac{\partial}{\partial x}\left(\frac{\partial u}{\partial y}\right) = -\frac{1}{2}(x^2 - y^2)^{-3/2}(2x)$$

$$= -\frac{x}{(x^2 - y^2)^{3/2}}$$

$$\therefore \quad \frac{\partial}{\partial x}\left(\frac{\partial u}{\partial y}\right) = -\frac{x}{(x^2 - y^2)^{3/2}}$$

$$\therefore \quad \frac{\partial^2 u}{\partial x\, \partial y} = \frac{-x}{(x^2 - y^2)^{3/2}} \qquad \ldots \text{(v)}$$

From equation (iv) and (v), we observe that

$$\frac{\partial^2 u}{\partial x\, \partial y} = \frac{\partial^2 u}{\partial y\, \partial x}$$

(ii)　We have,　　$u = \log\left(\dfrac{x^2 + y^2}{xy}\right)$ 　　　$\ldots$ (i)

$$\therefore \quad \frac{\partial u}{\partial x} = \frac{xy}{x^2 + y^2} \cdot \frac{\partial}{\partial x}\left(\frac{x^2 + y^2}{xy}\right)$$

$$= \frac{xy}{x^2 + y^2}\left[\frac{xy(2x) - (x^2 + y^2)y}{x^2 y^2}\right]$$

$$= \frac{2x^2 y - x^2 y - y^3}{xy(x^2 + y^2)} = \frac{x^2 y - y^3}{xy(x^2 + y^2)}$$

$$= \frac{x^2 - y^2}{x(x^2 + y^2)}$$

$$\therefore \quad \frac{\partial u}{\partial x} = \frac{x^2 - y^2}{x(x^2 + y^2)} \qquad \ldots \text{(ii)}$$

and
$$\frac{\partial u}{\partial y} = \frac{xy}{x^2 + y^2} \frac{\partial}{\partial y}\left(\frac{x^2 + y^2}{xy}\right)$$

$$= \frac{xy}{x^2 + y^2}\left[\frac{xy \cdot (2y) - (x^2 + y^2) \cdot x}{x^2 y^2}\right]$$

$$= \frac{2xy^2 - x^3 - xy^2}{xy(x^2 + y^2)}$$

$$= \frac{xy^2 - x^3}{xy(x^2 + y^2)} = \frac{y^2 - x^2}{y(x^2 + y^2)}$$

$$\therefore \quad \frac{\partial u}{\partial y} = -\frac{x^2 - y^2}{y(x^2 + y^2)} \qquad \ldots \text{(iii)}$$

Now differentiating equation (ii), w.r.t. y partially, we get

$$\frac{\partial}{\partial y}\left(\frac{\partial u}{\partial x}\right) = \frac{x(x^2 + y^2)(-2y) - (x^2 - y^2) \cdot (2xy)}{x^2 (x^2 + y^2)^2}$$

$$= \frac{-2xy\,[(x^2 + y^2) + (x^2 - y^2)]}{x^2 (x^2 + y^2)^2}$$

$$= \frac{-2xy(2x^2)}{x^2 (x^2 + y^2)^2} = -\frac{4xy}{(x^2 + y^2)^2}$$

$$\therefore \quad \frac{\partial}{\partial y}\left(\frac{\partial u}{\partial x}\right) = -\frac{4xy}{(x^2 + y^2)^2}$$

$$\therefore \quad \frac{\partial^2 u}{\partial y\, \partial x} = -\frac{4xy}{(x^2 + y^2)^2} \qquad \ldots \text{(iv)}$$

Now differentiating equation (iii) w.r.t. x partially, we get,

$$\frac{\partial}{\partial x}\left(\frac{\partial u}{\partial y}\right) = -\left[\frac{y(x^2 + y^2)(2x) - (x^2 - y^2)(2xy)}{y^2(x^2 + y^2)^2}\right]$$

$$= -\left[\frac{2yx^3 + 2xy^3 - 2yx^3 + 2xy^2}{y^2(x^2 + y^2)^2}\right]$$

$$= -\left[\frac{4xy^3}{y^2(x^2 + y^2)^2}\right] = -\frac{4xy}{(x^2 + y^2)^2}$$

$$\therefore \quad \frac{\partial}{\partial x}\left(\frac{\partial u}{\partial y}\right) = -\frac{4xy}{(x^2 + y^2)^2} \qquad \ldots \text{(v)}$$

$$\therefore \quad \frac{\partial^2 u}{\partial x\, \partial y} = -\frac{4xy}{(x^2 + y^2)^2} \qquad \ldots \text{(vi)}$$

From equation (iv) and (v), we observe that

$$\frac{\partial^2 u}{\partial x\, \partial y} = \frac{\partial^2 u}{\partial y\, \partial x}$$

3.1.3 Differential Equations, General Solution of Differential Equations

Definition : An equation involving dependent variables, independent variables and derivatives of dependent variables w.r.t. independent variables or function of this is called differential equation.

For example :

(i) $e^x\, dx + e^y\, dy = 0$

(ii) $y = x\,\dfrac{dy}{dx}$

(iii) $\left[1 + \left(\dfrac{dy}{dx}\right)^2\right]^{3/2} = c\,\dfrac{d^2y}{dx^2}$

(iv) $\dfrac{\partial^2 y}{\partial x^2} = \dfrac{1}{c}\,\dfrac{\partial^2 y}{\partial t^2}$

are all differential equations.

Definition : (Ordinary Differential Equation)

A differential equation which contains only one independent variable is called an ordinary differential equation i.e. A differential equation which contains only ordinary derivatives is called an ordinary differential equation.

For example :

(i) $1 + \left(\dfrac{dy}{dx}\right)^2 = \dfrac{d^3y}{dx^3}$

(ii) $\dfrac{d^2y}{dx^2} + 5\,\dfrac{dy}{dx} + 6y = 0$

(iii) $x\, dx + y\, dy = 0$

are all ordinary differential equations as they have only one independent variable x.

Definition : (Partial Differential Equation)

A differential equation which contains more than one independent variable is called **partial differential equation** i.e. A differential equation which contains partial derivatives is called a **partial differential equation.**

For example :

(i) $\dfrac{\partial^2 u}{\partial x^2} + \dfrac{\partial^2 u}{\partial y^2} + \dfrac{\partial^2 u}{\partial z^2} = 0$

(ii) $\dfrac{\partial^2 u}{\partial x^2} = \dfrac{1}{c}\dfrac{\partial^2 u}{\partial y^2}$

(iii) $x\dfrac{\partial u}{\partial x} + y\dfrac{\partial u}{\partial y} + z\dfrac{\partial u}{\partial z} = xyz$

are partial differential equations as they have more than one independent variable such as x, y, z.

Definition : (Order of Differential Equation)

The order of a differential equation is the order of highest order derivative appearing in it, provided the differential equation is free from integral sign.

Definition : (Degree of Differential Equation)

The degree of differential equation is the degree of the highest order derivative occurring in it, after the equation has been expressed in a form such that it is free from radicals and fractions as far as the derivatives are concerned. Also the derivatives occurring in the equation should not occur in the denominator and under radical sign.

Illustration 3.2 : Determine the order and degree of differential equation.

$$x\left(\frac{d^2 y}{dx^2}\right)^3 + y\left(\frac{dy}{dx}\right)^4 + y^4 = 0$$

Solution : The given differential equation contains second order derivative which is the highest order derivative, so the order of the differential equation is 2. The power of second order derivative is 3. So the degree of the differential equation is 3.

Hence, Order = 2 and Degree = 3.

Illustration 3.3 : Determine the order and degree of the differential equation.

$$\left[1 + \left(\frac{dy}{dx}\right)^2\right]^{1/2} = \left(\frac{d^2 y}{dx^2}\right)^{1/3}$$

Solution : The given differential equation contains radicals. First, we make the differential coefficients free from radicals.

For this, taking 6th power on both the sides of given differential equation, we get

$$\left[1+\left(\frac{dy}{dx}\right)^2\right]^3 = \left(\frac{d^2y}{dx^2}\right)^2$$

Now in this equation, the order of highest order derivative is 2. So its order is 2. The power of the highest derivative is 2, so its degree is 2.

Hence the given differential equation is of order 2 and degree 2.

Remark : If the given differential equation is not reduced to a polynomial in $\frac{dy}{dx}$, then its degree is not defined.

Solution of the Differential Equation

Definition : The solution of a differential equation is a function, which when substituted in the differential equation reduces it to an identity i.e. a solution of a differential equation is a relation between the dependent and independent variables such that, this relation and the values of derivatives calculated from it satisfy the differential equation.

For example : (i) Consider differential equation $\frac{dy}{dx} = \cos x$ and its solution $y = \sin x + c$.

Since, $$y = \sin x + c$$

$$\Rightarrow \qquad \frac{dy}{dx} = \cos x$$

Substituting the value of $\frac{dy}{dx}$ in the given differential equation, we get an identity $\cos x = \cos x$, i.e. $y = \sin x + c$ satisfies the given differential equation. Hence it is a solution of the differential equation.

Thus, the solution of differential equation is a function which when substituted in the differential equation reduces to an identity.

Illustration 3.4 : Verify that $y = A \cos x - B \sin x$ is a solution of the differential equation

$$\frac{d^2y}{dx^2} + y = 0$$

Solution : We have, $\qquad y = A \cos x - B \sin x \qquad\qquad\qquad …(i)$

Differentiating equation (i), w.r.t. x, we get

$$\frac{dy}{dx} = -A \sin x - B \cos x$$

Differentiating again w.r.t x, we get

$$\frac{d^2y}{dx^2} = -A \cos x + B \sin x$$

Substituting values of $\frac{d^2y}{dx^2}$ and y in L.H.S. of given differential equation, we get

$$\text{L.H.S.} = \frac{d^2y}{dx^2} + y$$

$$= -A \cos x + B \sin x = A \cos x - B \sin x$$
$$= 0 = \text{R.H.S.}$$

Thus, y = A cos x − B sin x satisfies given differential equation. Hence it is a solution of it.

General Solution of Differential Equation :

Definition : A general solution of an ordinary differential equation is a relation between dependent and independent variables in which the number of independent arbitrary constants involved in it is equal to the order of the equation.

The general solution is also called the **complete integral** or **complete primitive**.

Particular Solution :

Definition : Any solution obtained from the general solution by assigning particular values to the arbitrary constants involved in it is called a **particular solution** or **particular integral**.

Singular Solution :

Definition : The solution which can not be obtained from general solution by assigning particular values to the arbitrary constants, is called **singular solution**.

For example, consider the differential equation

$$\frac{d^2y}{dx^2} + 9y = 0$$

It can be shown that y = cos 3x is a solution of the differential equation.

Similarly, y = sin 3x is also a solution of the differential equation.

Further it can be shown that $y = A \cos 3x$, where A is arbitrary constant, is also a solution of the given equation.

Since the solution $y = \cos 3x$ is obtainable from the solution $y = A \cos 3x$ by giving A the value 1. Hence the solution $y = A \cos 3x$ is more general solution than the solution $y = \cos 3x$.

Similarly $y = B \sin 3x$, where B is arbitrary constant is also a more general solution than the solution $y = \sin 3x$.

It can be shown that $y = A \cos 3x + B \sin 3x$, where A and B are arbitrary constants is a still more general solution of the given differential equation, than all the previous solutions viz $y = \cos 3x$, $y = A \cos 3x$, $y = \sin 3x$ and $y = B \sin 3x$, because all these solutions are obtained from $y = A \cos 3x + B \sin 3x$ by giving particular values to arbitrary constants A and B.

Number of arbitrary constants in solution $y = A \cos 3x + B \sin 3x$ is two and order of given differential equation is two. Thus, the number of arbitrary constants is equal to order of given differential equation. Hence,

$$y = A \cos 3x + B \sin x \qquad \qquad \text{... (i)}$$

is a general solution of the given differential equations.

The solutions $y = \cos 3x$, $y = A \cos 3x$, $y = \sin 3x$ and $y = B \sin 3x$ are particular solutions of the equation, because these solutions are obtained by giving particular values to the arbitrary constants in the general solution $y = A \cos 3x + B \sin 3x$.

Formation of Differential Equation :

The general solution of a first order differential equation contains one arbitrary constant, which is called a **parameter**. If this parameter takes various values, then we get a family of curves of one parameter. For example, The equation $x^2 + y^2 = c^2$ represent one parameter family of circles, if c takes all real values.

Suppose there is an equation representing a family of curves (general solution) containing n arbitrary constants. Then in order to find its differential equation we proceed as follows :

Step - 1 : Differentiate given equation of family of curves n-times to get n more equations containing n arbitrary constants and derivatives.

Step - 2 : Eliminate all the n-constants from all these (n + 1) equations to get an equation containing n^{th} order derivative, which is the required differential equation of the given family of curves.

Illustrative Examples

Example 3.5 : *Find the differential equation of the family of all straight lines passing through the origin.*

Solution : The general equation of the family of all straight lines passing through the origin is

$$y = mx \qquad \ldots \text{(i)}$$

where, m is an arbitrary constant (parameter)

Equation (i) contains one arbitrary constant. Therefore, required differential equation should be of order one.

Differentiating, equation (i) w.r.t. x, we get

$$\frac{dy}{dx} = m \qquad \ldots \text{(ii)}$$

Eliminating m between equations (i) and (ii), we get

$$y = x\frac{dy}{dx}$$

which is the required differential equation.

Example 3.6 : *Form the differential equation from*

$$y = a \cos x + b \sin x$$

where, a and b are arbitrary constants.

Solution : We have to form an differential equation whose solution is

$$y = a \cos x + b \sin x \qquad \ldots \text{(i)}$$

The given solution of equation (i) contains two arbitrary constants. Therefore required differential equation should be of second order.

To obtain the differential equation whose solution is (i). We have to differentiate equation (i) two times.

$\therefore$ Differentiating equation (i) w.r.t. x, we get

$$\frac{dy}{dx} = -a \sin x + b \cos x \qquad \ldots \text{(ii)}$$

Differentiating again w.r.t. x, we get

$$\frac{d^2y}{dx^2} = -a \cos x - b \sin x$$

$$= -(a \cos x + b \sin x)$$

$$= -y$$

$$\therefore \qquad \frac{d^2y}{dx^2} = -y \qquad \qquad \text{... (iii)}$$

Eliminating a and b between equations (i), (ii) and (iii), we get

$$\frac{d^2y}{dx^2} + y = 0$$

which is the required differential equation.

Example 3.7 : *Determine order and degree of the differential equation*

$$x\frac{dy}{dx} + \frac{2}{\frac{dy}{dx}} = y^2$$

Solution : The given differential equation contains the derivative in the denominator. First we make the differential equation, so that derivative should not occur in denominator.

For this multiplying both the sides of given equation by $\frac{dy}{dx}$, we get

$$x\left(\frac{dy}{dx}\right)^2 + 2 = y^2\frac{dy}{dx}$$

Now in this equation, the highest order derive is $\frac{dy}{dx}$, which is of order 1. So the order of given differential equation is 1. The power of highest order derivative $\frac{dy}{dx}$ is 2, so its degree is 2.

Hence, order = 1 and degree = 2.

Example 3.8 : *Determine the order and degree of the differential equation*

$$y = px + \sqrt{a^2p^2 + b^2},$$

where, $\qquad p = \dfrac{dy}{dx}$

Solution : The given differential equation is

$$y = px + \sqrt{a^2p^2 + b^2},$$

where, $\qquad p = \dfrac{dy}{dx}$

Making the given differential coefficients free from radicals, we get

$$\Rightarrow \qquad y - px = \sqrt{ap^2 + b^2}$$

$$\Rightarrow \qquad (y - px)^2 = a^2 p^2 + b^2$$

$$\text{[Squaring both sides]}$$

$$\Rightarrow \qquad y^2 + p^2 x^2 - 2y\,px = a^2 p^2 + b^2$$

$$\Rightarrow \qquad (x^2 - a^2)\,p^2 - 2xyp + y^2 - b^2 = 0$$

$$\text{i.e.} \quad (x^2 - a^2)\left(\frac{dy}{dx}\right)^2 - 2xy\,\frac{dy}{dx} + y^2 - b^2 = 0 \qquad \because\ p = \frac{dy}{dx}$$

Now in this equation, the order of highest derivative is one and its power is 2.

Hence the given differential equation is of order 1 and degree 2.

Example 3.9 : *Determine the order and degree of differential equation*

$$\frac{d^2 y}{dx^2} + sin\left(\frac{dy}{dx}\right) = 0$$

Solution : We have, $\quad \dfrac{d^2 y}{dx^2} + sin\left(\dfrac{dy}{dx}\right) = 0$

This differential equation contains second order derivative term which is the highest order derivative in it, so its order is 2, but it can not be reduced to a polynomial in $\dfrac{dy}{dx}$, therefore its degree is not defined.

Example 3.10 : *Find the differential equation of the family of curves*

$$y^2 - 2ay + x^2 = a^2$$

where, a is arbitrary constant.

Solution : We have, $\quad y^2 - 2ay + x^2 = a^2 \qquad \ldots \text{(i)}$

where, a is an arbitrary constant.

The equation (i) contains one arbitrary constant. Therefore, required differential equation should be of order one.

Differentiating equation (i) w.r.t. x, we get

$$2y\,\frac{dy}{dx} - 2a\,\frac{dy}{dx} + 2x = 0$$

$$\therefore \qquad a = y + \frac{x}{\left(\dfrac{dy}{dx}\right)} \qquad \ldots \text{(ii)}$$

Eliminating a between equation (i) and (ii), we get

$$y^2 - 2\left[y + \frac{x}{\left(\frac{dy}{dx}\right)}\right]y + x^2 = \left[y + \frac{x}{\left(\frac{dy}{dx}\right)}\right]^2$$

$$y^2 - 2\left[y + \frac{x}{\left(\frac{dy}{dx}\right)}\right]y + x^2 = y^2 + \frac{x^2}{\left(\frac{dy}{dx}\right)^2} + \frac{2xy}{\left(\frac{dy}{dx}\right)}$$

or $\quad -2y\frac{dy}{dx}\left[y\frac{dy}{dx} + x\right] + x^2\left(\frac{dy}{dx}\right)^2 = x^2 + 2xy\frac{dy}{dx}$

or $\quad (x^2 - 2y^2)\left(\frac{dy}{dx}\right)^2 - 4xy\frac{dy}{dx} - x^2 = 0$

which is the required differential equation.

Example 3.11 : *Find the differential equation of the family of curves $y = \frac{a}{x} + b$; where a and b are arbitrary constants.*

Solution : We have, $y = \frac{a}{x} + b$ $\qquad$ ... (i)

where, a and b are arbitrary constants.

The equation (i) contains two arbitrary constants. Therefore required differential equation should be of second order.

Differentiating equation (i) w.r.t. x, we get

$$\frac{dy}{dx} = -\frac{a}{x^2} \Rightarrow x^2\frac{dy}{dx} = -a \qquad ... (i)$$

Differentiating again w.r.t x, we get

$$x^2\frac{d^2y}{dx^2} + 2x\frac{dy}{dx} = 0$$

or $\qquad x\frac{d^2y}{dx^2} + 2\frac{dy}{dx} = 0$

which is the required differential equation.

3.2 Separable Equations

3.2.1 Variable Separable Equation

The most general differential equation of first order and first degree is of the form

$$M + N\frac{dy}{dx} = 0 \qquad ... (i)$$

where, M and N are functions of x and y.

The equation (i) can be rewritten by using differentials as

$$M\ dx + N\ dy\ =\ 0 \qquad\qquad \text{... (ii)}$$

There are various types of differential equations of first order and first degree. Type of the differential equation (i) and hence (ii) is determined from the nature of functions M and N.

Now will see different types of differential equations of first order and first degree and we will see how to solve these equations.

Type - 1 : Variables Separable Type :

The differential equation

$$M\ dx + N\ dy\ =\ 0 \qquad\qquad \text{... (i)}$$

can be solved, if M is function of x alone and N is function of y alone.

Let $\qquad\qquad M\ =\ f(x)$ and $N = g(y),\ $ say

Then, equation (i) becomes,

$$f(x)\ dx + g(y)\ dy\ =\ 0\ \text{ i.e. }\ g(y)\ y' + f(x) = 0 \qquad \text{... (ii)}$$

Now, suppose that $g(y)y' = f(x)$ is a variable separable equation. Assume that y is its solution. Let $G(y)$ and $F(x)$ be anti-derivatives of $g(y)$ and $f(x)$ respectively so that

$$\int g(y)\ dy\ =\ G(y) \text{ and } \int f(x)\ dx = F(x)$$

Then $\qquad\qquad G'(y)\ =\ g(y)$ and $F'(x) = f(x)$

$$\therefore \qquad\qquad \frac{d}{dx}\,G(y(x))\ =\ G'(y)\cdot y'(x) = g(y)\ y'(x)$$

$\therefore\quad$ From given equation,

$$\frac{d}{dx}\,(G(y(x)))\ =\ \frac{d}{dx}\,F(x)$$

Integrating, this equation, we get

$$G(y(x))\ =\ F(x) + c,$$

the general solution of the given equation.

Hence, the general solution of the equation

$$g(y)\,\frac{dy}{dx}\ =\ f(x) \text{ is } \int g(y)\ dy = \int f(x)\ dx + c,$$

where, c is arbitrary constant of integration.

This solution contains one arbitrary constant and our given differential equation is of order one. Hence it is a general solution of equation (i).

In this equation (ii), variables x and y occur in separate terms i.e. first term containing function of variable x and its differential dx and second term containing function of variable y and its differential dy. Hence such a differential equation is called **"variable separable type"** and in this case we say that variables are separated.

Thus, if required, adjust the equation so as to get a separate variable type and then the general solution is obtained by integrating the equation term by term.

Illustrative Examples

Example 3.12 : *Solve the initial value problems :*

(i) $y' = -\dfrac{x}{y}, \; y(1) = 1$ (ii) $y' = -\dfrac{x}{y}, \; y(1) = -2$

Solution : (i) Separating the variables, we get

$$\text{ydy} = -\text{ xdx} \;\Rightarrow\; \int \text{ydy} = \int -\text{xdx} + c \;\Rightarrow\; \frac{y^2}{2} = \frac{-x^2}{2} + c$$

$$\Rightarrow\; y^2 + x^2 = c_1, \; \text{where } c_1 = 2c.$$

Now by the initial condition, $y = 1$, when $x = 1$, put in the solution, we get $1^2 + 1^2 = c_1 \;\Rightarrow\; c_1 = 2$.

∴   The particular solution is $x^2 + y^2 = 2$.

(ii)   We know that the general solution is

$$x^2 + y^2 = c_1.$$

Now, by the initial condition $y = -2$ when $x = 1$, we get $c_1 = 5$, so that the particular solution is $x^2 + y^2 = 5$.

Note that if $-\sqrt{2} < x < \sqrt{2}$, then the solution (1) is

$$y = \sqrt{2 - x^2}.$$

This gives an explicit formula for y in terms of x.

Theorem 2 Suppose that the function $f(x)$, $g(y)$ are continuous on the intervals (a, b) and (c, d) respectively.

Let $F(x) = \int f(x)\, dx$, $G(y) = \int g(y)\, dy$ and let $y_0 \in (c, d)$ such that $g(y_0) \neq 0$ and $c = G(y_0) - F(x_0)$.

Then there is a function $y = y(x)$ defined on some open interval (a_1, b_1), where $a \leq a_1 < x_0 < b_1 \leq b$, such that $y(x_0) = y_0$ and

$$G(y) \;=\; F(x) + c \text{ for } a_1 < x < b. \qquad \ldots (1)$$

Here y is a solution of the initial value problem

$$g(y)\, y' \;=\; f(x), \quad y(x_0) = y_0 \qquad\qquad \dots (2)$$

Equation (1) is called as an implicit solution of equation (2). ∎

Example 3.13 : *Find an implicit solution of*

$$y' = \frac{2x + 1}{5y^4 + 1}, \quad y(2) = 1.$$

Solution : By separating variables, we get the general solution

$$y^5 + y \;=\; x^2 + x + c$$

Now, using the initial condition $y = 1$, when $x = 1$, we get $c = -4$. Therefore, the solution is

$$y^5 + y \;=\; x^2 + x - 4$$

Example 3.14 : *Solve,* $x^3\, dx + (y + 1)^2\, dy = 0$

Solution : Here the variables are separated. Hence integrating term by term, we get

$$\int x^3\, dx + \int (y + 1)^2\, dy \;=\; c_1$$

i.e.
$$\frac{x^4}{4} + \frac{(y + 1)^3}{3} \;=\; c_1$$

i.e.
$$3x^4 + 4(y + 1)^3 \;=\; 12\, c_1$$

$$\therefore \quad 3x^4 + 4(y + 1)^3 \;=\; c, \quad \text{where,} \qquad c = 12\, c_1$$

which is the required general solution.

Example 3.15 : *Solve,* $\dfrac{dy}{dx} = xy + x + y + 1$

Solution : We have,
$$\frac{dy}{dx} \;=\; xy + x + y + 1$$

$$\Rightarrow \qquad \frac{dy}{dx} \;=\; (1 + x)\,(1 + y)$$

Separating the variables, we get

$$\frac{dy}{1 + y} \;=\; (1 + x)\, dx$$

Integrating, we get
$$\log (1 + y) \;=\; x + \frac{x^2}{2} + c$$

which is the required general solution.

Differential Equations Reducible to Variable Separable Form :

Some times, we come across with differential equations, in which the variables can not be separated. In such type of differential equation, some suitable substitution reduces it to a form in which the variables are separated.

The equations of the type

$$\frac{dy}{dx} = \phi(ax + by), \quad \frac{dy}{dx} = f(ax + by + c) \text{ and } \frac{dy}{dx} = f\left(\frac{y}{x}\right)$$

can be transformed to variable separable type by the substitution

$ax + by = v, \ ax + by + c = v \text{ and } \frac{y}{x} = v$ respectively.

Illustrative Examples

Example 3.16 : *Solve,* $\dfrac{dy}{dx} = sin\ (x + y)$

Solution : We have, $\qquad \dfrac{dy}{dx} = sin\ (x + y)$ $\qquad$ ... (i)

Put $x + y = v$, so that $1 + \dfrac{dy}{dx} = \dfrac{dv}{dx}$ $\quad \therefore \ \dfrac{dy}{dx} = \dfrac{dv}{dx} - 1$

By this substitution given differential equation (i) reduces to

$$\frac{dv}{dx} - 1 = sin\ v$$

$$\Rightarrow \qquad \frac{dv}{dx} = 1 + sin\ v$$

Separating the variables, we get

$$\frac{dv}{1 + sin\ v} = dx$$

Integrating both the sides, we get

$$\int \frac{dv}{1 + sin\ v} = \int dx + c$$

$$\Rightarrow \qquad \int \frac{(1 - sin\ v)}{(1 + sin\ v)\,(1 - sin\ v)}\, dv = x + c$$

$$\Rightarrow \qquad \int \frac{1 - sin\ v}{1 - sin^2 v}\, dv = x + c$$

$$\Rightarrow \qquad \int \frac{1 - sin\ v}{cos^2\ v}\, dv = x + c$$

$$\Rightarrow \qquad \int \left[\frac{1}{cos^2 v} - \frac{sin\ v}{cos^2 v} \right] dv = x + c$$

$$\Rightarrow \qquad \int [\sec^2 v - \sec v \tan v]\, dv = x + c$$

$$\Rightarrow \qquad \tan v - \sec v = x + c$$

$$\therefore \qquad \tan (x + y) - \sec (x + y) = x + c \qquad \because v = x + y$$

which is the required solution.

Example 3.17 : *Solve,* $\dfrac{dy}{dx} = (4x + y + 1)^2$

Solution : We have, $\dfrac{dy}{dx} = (4x + y + 1)^2 \qquad \ldots (i)$

Put $v = 4x + y + 1$, then $\dfrac{dv}{dx} = 4 + \dfrac{dy}{dx}$, so that $\dfrac{dy}{dx} = \dfrac{dv}{dx} - 4$

Substituting these values in equation (i), we get

$$\frac{dv}{dx} - 4 = v^2$$

$$\Rightarrow \qquad \frac{dv}{dx} = v^2 + 4$$

Separating the variables, we get

$$\frac{dv}{v^2 + 4} = dx$$

Integrating both the sides, we get

$$\int \frac{dv}{v^2 + 4} = \int dx + c$$

$$\Rightarrow \qquad \frac{1}{2} \tan^{-1}\left(\frac{v}{2}\right) = x + c$$

or $\qquad 4x + y + 1 = 2 \tan (2x + 2c) \qquad \because v = 4x + y + 1$

which is a required solution.

3.3 Transformation of Non-Linear Equations to Separable Equations

3.3.1 Homogeneous Differential Equations

Definition : The differential equation $M\, dx + N\, dy = 0$ is called homogeneous differential equation, if M and N are homogeneous functions of the same degree.

Such a homogeneous differential equation can be written as

$$\frac{dy}{dx} = \frac{f_1(x,\, y)}{f_2(x,\, y)}$$

where $f_1(x,\, y)$ and $f_2(x,\, y)$ are **homogeneous functions of same degree**.

Theorem 3 When $M\,dx + N\,dy = 0$ is homogeneous differential equation, the substitution $y = vx$ will separate the variables.

Proof : Let, $\quad M\,dx + N\,dy = 0 \qquad\qquad\qquad …\,(i)$

be the homogeneous differential equation. This equation can be rewritten as

$$\frac{dy}{dx} = \frac{-M}{N}$$

Let $-M = f_1(x, y)$, $N = f_2(x, y)$, where $f_1(x, y)$ and $f_2(x, y)$ are homogeneous functions of the same degree. Let it be of degree n.

$$\therefore \qquad \frac{dy}{dx} = \frac{f_1(x, y)}{f_2(x, y)} \qquad\qquad …\,(ii)$$

Now we show that the substitution $y = vx$ transforms (ii) and hence (i) to separate variable type.

Putting $y = vx$, so that $\dfrac{dy}{dx} = v + x\dfrac{dv}{dx}$ in equation (ii).

Equation (ii) transforms to

$$v + x\frac{dv}{dx} = \frac{f_1(x, vx)}{f_2(x, vx)}$$

$$= \frac{x^n\,\phi_1(v)}{x^n\,\phi_2(v)}$$

$$\because\ f_1 \text{ and } f_2 \text{ are homogeneous functions of degree } n$$

$$\therefore \qquad f_1(x, y) = x^n\,\phi_1\left(\frac{y}{x}\right)$$

$$= x^n\,\phi_1(v) \qquad\qquad \because y = vx$$

$$\text{and} \qquad f_2(x, y) = x^n\,\phi_2(v)$$

$$v + x\frac{dv}{dx} = \frac{\phi_1(v)}{\phi_2(v)} = \phi(v)\ \text{(say)}$$

$$\therefore \qquad v + x\frac{dv}{dx} = \phi(v)$$

$$\therefore \qquad x\frac{dv}{dx} = \phi(v) - v$$

$$\Rightarrow \qquad \frac{dv}{\phi(v) - v} = \frac{dx}{x}$$

Thus the variables are separated.

Integrating on both the sides, we get

$$\int \frac{dv}{\phi(v) - v} = \int \frac{dx}{x} + c$$

$$\Rightarrow \qquad \int \frac{dv}{\phi(v)} = \log (x) + c$$

Resubstituting $v = \dfrac{y}{x}$, we get the solution. ■

Method to Solve Homogeneous Differential Equation :

Homogeneous differential equation can always be solved by substituting $y = vx$. It reduces to variable separable type and variable separable type differential equation can be solved by term and term integration.

Illustrative Examples

Example 3.18 : *Solve,* $(x^2 + xy) \dfrac{dy}{dx} = 2xy$

Solution : The given differential equation can be written as

$$\frac{dy}{dx} = \frac{2xy}{x^2 + xy} \qquad \qquad \ldots \text{(i)}$$

Clearly equation (i) is a homogeneous differential equation.

Putting $y = vx$ $\quad \therefore \dfrac{dy}{dx} = v + x \dfrac{dv}{dx}$ in equation (i), we get

$$v + x \frac{dv}{dx} = \frac{2vx^2}{x^2 + vx^2}$$

$$\therefore \qquad v + x \frac{dv}{dx} = \frac{2v}{1 + v}$$

$$\Rightarrow \qquad x \frac{dv}{dx} = \frac{2v}{1 + v} - v$$

$$= \frac{v - v^2}{1 + v}$$

$$\Rightarrow \qquad x \frac{dv}{dx} = - \frac{v^2 - v}{v + 1}$$

Separating the variables, we get

$$\frac{v + 1}{v^2 - v} dv = - \frac{dx}{x}$$

Integrating both the sides, we get

$$\int \frac{v+1}{v^2 - v} \, dv = - \int \frac{dx}{x} + c_1$$

$$= - \log(x) + c_1$$

$$\Rightarrow \qquad \int \frac{v+1}{v(v-1)} \, dv = - \log(x) + c_1$$

$$\Rightarrow \qquad \int \left[\frac{2}{v-1} - \frac{1}{v} \right] dv = - \log(x) + c_1$$

$$\Rightarrow \qquad 2 \log(v-1) - \log(v) = - \log(x) + \log c$$

$$\Rightarrow \qquad \log(x) + 2 \log(v-1) = - \log(v) = \log c$$

$$\Rightarrow \qquad \log\left(\frac{x(v-1)^2}{v} \right) = \log c$$

$$\Rightarrow \qquad \left(\frac{x(v-1)^2}{v} \right) = c$$

$$\Rightarrow \qquad \frac{x(v-1)^2}{v} = c$$

$$\Rightarrow \qquad x(v-1)^2 = cv$$

$$\Rightarrow \qquad x\left(\frac{y}{x} - 1 \right)^2 = c\left(\frac{y}{x} \right)$$

$$\Rightarrow \qquad (y-x)^2 = cy$$

which is required solution.

Example 3.19 : *Solve,* $(x - y)\, dx + (x + y)\, dy = 0$

Solution : The given differential equation can be written as

$$\frac{dy}{dx} = \frac{y-x}{y+x} \qquad\qquad \ldots \text{(i)}$$

Clearly equation (i) is homogeneous differential equation.

Putting $y = vx$, so that $\dfrac{dy}{dx} = v + x \dfrac{dv}{dx}$ in equation (i), we get

$$v + x\frac{dv}{dx} = \frac{vx - x}{vx + x}$$

$$\Rightarrow \qquad v + x\frac{dv}{dx} = \frac{v-1}{v+1}$$

$$\Rightarrow \qquad x\frac{dv}{dx} = \frac{v-1}{v+1} - v$$

$$\Rightarrow \qquad x\frac{dv}{dx} = \frac{-1-v^2}{v+1} \quad \Rightarrow \quad x\frac{dv}{dx} = -\frac{v^2+1}{v+1}$$

Separating the variables, we get

$$\frac{v+1}{v^2+1}\,dv = \frac{-dx}{x}$$

$$\Rightarrow \qquad \frac{v+1}{v^2+1}\,dv + \frac{dx}{x} = 0$$

Integrating both the sides, we get

$$\int \frac{v+1}{v^2+1}\,dv + \int \frac{dx}{x} = c$$

$$\Rightarrow \quad \int \left[\frac{v}{v^2+1} + \frac{1}{v^2+1} \right] dv + \int \frac{dx}{x} = c$$

$$\Rightarrow \quad \frac{1}{2} \int \frac{2v}{v^2+1}\,dv + \int \frac{dv}{v^2+1} + \int \frac{dx}{x} = c$$

$$\Rightarrow \quad \frac{1}{2} \log (v^2+1) + \tan^{-1} v + \log (x) = c$$

$$\Rightarrow \quad \frac{1}{2} \log \left(\frac{y^2}{x^2}+1 \right) + \log (x) + \tan^{-1} \left(\frac{y}{x} \right) = c$$

$$\Rightarrow \quad \frac{1}{2} \log \left(\frac{x^2+y^2}{x^2} \right) + \log (x) + \tan^{-1} \left(\frac{y}{x} \right) = c$$

$$\Rightarrow \quad \frac{1}{2} \log (x^2+y^2) - \frac{1}{2} \log (x^2) + \log (x) + \tan^{-1} \left(\frac{y}{x} \right) = c$$

$$\Rightarrow \quad \frac{1}{2} \log (x^2+y^2) - \log (x) + \log (x) + \tan^{-1} \left(\frac{y}{x} \right) = c$$

$$\Rightarrow \quad \frac{1}{2} \log (x^2+y^2) + \tan^{-1} \left(\frac{y}{x} \right) = c$$

which is the required solution.

3.3.2 Equation Reducible to Homogeneous Form

The differential equation of the type

$$\frac{dy}{dx} = \frac{a_1x + b_1y + c_1}{a_2x + b_2y + c_2} \qquad \qquad \ldots \text{(i)}$$

can be reduced to the homogeneous form as follows –

We transform the variables x and y to new variables X and Y by writing

$$x = X + h \ \text{ and } \ y = Y + k \qquad \qquad \ldots \text{(ii)}$$

where, h and k are constants to be suitably determined.

From equation (ii), we get $dx = dX$, $dy = dY$

$\therefore$ The given equation (i) becomes

$$\frac{dY}{dX} = \frac{a_1(X + h) + b_1(Y + k) + c_1}{a_2(X + h) + b_2(Y + k) + c_2}$$

$$\Rightarrow \qquad \frac{dY}{dX} = \frac{a_1X + b_1Y + (a_1h + b_1k + c_1)}{a_2X + b_2Y + (a_2h + b_2k + c_2)} \qquad \cdots \text{(iii)}$$

We now determine constants h and k such that

$$a_1h + b_1k + c_1 = 0 \text{ and } a_2h + b_2k + c_2 = 0 \qquad \cdots \text{(iv)}$$

Solving equations (iv) for h and k, we get

$$h = \frac{b_1c_2 - b_2c_1}{a_1b_2 - a_2b_1} , \quad k = \frac{c_1a_2 - c_2a_1}{a_1b_2 - a_2b_1} \qquad \cdots \text{(v)}$$

In view of equation (iv), equation (iii), becomes

$$\frac{dY}{dX} = \frac{a_1X + b_1Y}{a_2X + b_2Y} \qquad \cdots \text{(vi)}$$

This is a homogeneous differential equation of first order and first degree in X and Y. It can therefore, be solved by substitution $Y = vX$.

Let the solution of equation (vi) be

$$f(X, Y) = c$$

Resubstituting $X = x - h$, $Y = y - k$, we get the solution of the given equation as

$$f(x - h, y - k) = c$$

Working Rule : To solve $\dfrac{dy}{dx} = \dfrac{a_1x + b_1y + c_1}{a_2x + b_2y + c_2}$:

(A) When $\dfrac{a_1}{a_2} \neq \dfrac{b_1}{b_2}$:

(i)   Put $x = X + h$, $y = Y + k$ in the given differential equation.

(ii) Equate the constant terms in the numerator and denominator to zero and find the values of h and k.

(iii) Solve the resulting equation in X and Y.

(iv) Replace X by $x - h$ and Y by $y - k$.

(v) Put the value of h and k.

(B) When $\dfrac{a_1}{a_2} = \dfrac{b_1}{b_2}$:

(i)   Put $a_1x + b_1y = v$

(ii) Solve the resulting equation.

Illustrative Examples

Example 3.20 : *Solve, $(2x + y + 3)\, dx = (2y + x + 1)\, dy$*

Solution : The given differential equation can be written as

$$\frac{dy}{dx} = \frac{2x + y + 3}{2y + x + 1} \qquad \qquad \ldots \text{(i)}$$

This is the equation reducible to homogeneous form. Note that $\dfrac{a_1}{a_2} \neq \dfrac{b_1}{b_2}$

To solve it, put $x = X + h$, $y = Y + k$, so that $dx = dX$, $dy = dY$ in equation (i), we get

$$\frac{dY}{dX} = \frac{2X + Y + (2h + k + 3)}{2Y + X + (h + 2k + 1)} \qquad \qquad \ldots \text{(ii)}$$

Choosing h, k so that

$$2h + k + 3 = 0 \ \text{ and } \ h + 2k + 1 = 0$$

Solving these equations for h and k, we get

$$h = -\frac{5}{3}, \ k = \frac{1}{3}$$

With these values of h and k, equation (ii) becomes

$$\frac{dY}{dX} = \frac{2X + Y}{2Y + X} \qquad \qquad \ldots \text{(iii)}$$

which is the reduced homogeneous equation.

Now put $Y = vX$ so that $\dfrac{dY}{dX} = v + X\dfrac{dv}{dx}$ in equation (iii)

We get, $$v + X\frac{dv}{dX} = \frac{2X + vX}{2VX + X} = \frac{2 + v}{2v + 1}$$

$$\Rightarrow \qquad X\frac{dv}{dX} = \frac{v + 2}{2v + 1} - v$$

$$= \frac{-2(v^2 - 1)}{2v + 1}$$

$$\Rightarrow \qquad X\frac{dv}{dX} = -\frac{2(v^2 - 1)}{2v + 1}$$

Separating the variables, we get

$$\frac{2v + 1}{v^2 - 1}\, dv = -2\,\frac{dX}{X}$$

Integrating both the sides, we get

$$\int \frac{2v+1}{v^2-1}\, dv = -2\int \frac{dX}{X} + c_1$$

$$\Rightarrow \int \left[\frac{2v}{v^2-1} + \frac{1}{v^2-1}\right] dv = -2 \log (X) + c_1$$

$$\Rightarrow \log (v^2-1) + \frac{1}{2}\log\left(\frac{v-1}{v+1}\right) = -2\log(X) + c_1$$

$$\Rightarrow 2\log\left|v^2-1\right| + \log\left(\frac{v-1}{v+1}\right) = -4\log(X) + 2c_1$$

$$\Rightarrow \log\left((v^2-1)^2 \cdot \frac{v-1}{v+1}\, X^4\right) = \log c, \quad \text{where, } 2c_1 = \log c$$

$$\Rightarrow \log\left[(v-1)^3 (v+1)\, X^4\right] = \log c$$

$$\Rightarrow \left[(v-1)^3 (v+1)\, X^4\right] = c$$

$$\Rightarrow \left(\frac{Y}{X}-1\right)^3 \left(\frac{Y}{X}+1\right) X^4 = c$$

$$\Rightarrow \frac{(Y-X)^3}{X^3} \cdot \frac{(Y+X)}{X}\, X^4 = c$$

$$\Rightarrow (Y-X)^3 (Y+X) = c$$

$$\Rightarrow (y-x-2)^3 \left(x+y+\frac{4}{3}\right) = c$$

$$\because \quad X = x - h = x + \frac{5}{3} \ \text{ and } \ Y = y - k = y - \frac{1}{3}$$

which is the required solution.

Example 3.21 : *Solve,* $\dfrac{dy}{dx} = \dfrac{2x+9y-20}{6x+2y-10}$

Solution : The given differential equation is

$$\frac{dy}{dx} = \frac{2x+9y-20}{6x+2y-10} \qquad \ldots (i)$$

This is the equation reducible to homogeneous form.

To solve it, put $x = X + h$, $y = Y + k$, so that $dx = dX$ and $dy = dY$ in equation (i), we get

$$\frac{dY}{dX} = \frac{2X+9Y+(2h+9k-20)}{6X=2Y+(6h+2k-10)} \qquad \ldots (ii)$$

Choosing h and k, so that

$$2h+9k-20 = 0 \ \text{ and } \ 6h+2k-10 = 0$$

Solving these equations for h and k, we get

$$h = 1 \text{ and } k = 2$$

With these values of h and k, equation (ii) becomes

$$\frac{dY}{dX} = \frac{2X + 9Y}{6X + 2Y} \qquad \qquad \text{... (iii)}$$

which is the reduced homogeneous equation. To solve it, put $Y = vX$, so that

$$\frac{dY}{dX} = v + X\frac{dv}{dX} \text{ in equation (iii), we get}$$

$$v + X\frac{dv}{dX} = \frac{2X + 9vX}{6X + 2vX} = \frac{2 + 9v}{6 + 2v}$$

$$\Rightarrow \qquad X\frac{dv}{dX} = \frac{2 + 9v}{6 + 2v} - v$$

$$\Rightarrow \qquad X\frac{dv}{dx} = -\frac{2v^2 - 3v - 2}{xv + 6}$$

Separating the variables, we get

$$\frac{2v + 6}{2v^2 - 3v - 2}\,dv + \frac{dX}{X} = 0$$

Integrating, we get

$$\int \frac{2v + 6}{2v^2 - 3v - 2}\,dv + \int \frac{dX}{X} = c_1$$

$$\Rightarrow \qquad \int \left[\frac{2}{v - 2} - \frac{2}{2v + 1}\right]dv + \int \frac{dX}{X} = c_1$$

$$\Rightarrow \qquad 2\int \frac{dv}{v - 2} - 2\int \frac{1}{2v + 1}\,dv + \int \frac{dx}{X} = c_1$$

$$\Rightarrow \qquad 2\log(v - 2) - \log(2v + 1) + \log X = \log c$$

$$\Rightarrow \qquad \left(\frac{X(v - 2)^2}{2v + 1}\right) = \log c$$

$$\Rightarrow \qquad \frac{X(v - 2)^2}{2v + 1} = c$$

$$\Rightarrow \qquad \frac{X\left[\dfrac{Y}{X} - 2\right]^2}{2\left(\dfrac{y}{x}\right) + 1} = c$$

$$\Rightarrow \qquad \frac{(y - 2x)^2}{2X + X} = c$$

$$\Rightarrow \qquad \frac{(Y - 2X)^2}{2Y + X} = c$$

$$\Rightarrow \qquad (y - 2x)^2 = (x + 2y - 5)$$

$$\because X = x - 1 \text{ and } Y = y - 2$$

which is the required solution.

3.4 Linear First Order Differential Equations

Definition : A differential equation is said to be linear, when the dependent variable and its derivatives appears only in the first degree. No term of linear differential equation involves the product of any two derivatives or of the dependent variable and any derivative.

For example :

(i) $\dfrac{d^2y}{dx^2} + x\dfrac{dy}{dx} + y = 0$ is linear differential equation.

(ii) $\dfrac{d^2y}{dx^2} + xy^2 = 9$ is not a linear differential equation, since dependent variable y appears in second degree.

(iii) $x\dfrac{dy}{dx} + y = x$ is a linear differential equation.

In this text, we will consider linear differential equations of first order.

Linear Differential Equation of First Order :

Definition : Any differential equation which is of the form

$$\frac{dy}{dx} + Py = Q$$

where, P and Q are constants or functions of x only (and not of y) is called a *linear differential equation of first order in dependent variable y* and independent variable x.

Similarly, a differential equation

$$\frac{dx}{dy} + Px = Q$$

where, P and Q are constants or functions of y alone (and not of x) is called a *linear differential equation of first order in which the dependent variable is x* and independent variable is y.

Definition : Consider the linear differential equation $\dfrac{dy}{dx} + py = Q$.

If $Q(x) = 0$, then the linear equation $\dfrac{dy}{dx} + py = 0$ is called as homogeneous linear differential equation. If $Q(x) \neq 0$, then the equation $\dfrac{dy}{dx} + p(x)y = Q(x)$ is called as non-homogeneous linear equation.

3.4.1 Method of Solution of Linear Differential Equation of First Order $\dfrac{dy}{dx} + Py = Q$; where P and Q are functions of x only

Theorem 4 If $P(x)$ is continuous on (a, b) then the general solution of the homogeneous equation $y' + P(x)y = 0$ on (a, b) is

$$y = ce^{-\int P(x)\,dx}$$

Proof : If $y = ce^{-\int P(x)\,dx}$, differentiating w.r.t. x

we get, $\qquad y' = -P(x)\,ce^{-\int P(x)\,dx} = -P(x)\,y,$

so that $\qquad y' + P(x)\,y = 0$

$\therefore \quad y = ce^{-\int P(x)\,dx}$ is a solution of $y' + P(x)\,y = 0$.

Now, we will show that any solution can be written as $y = ce^{-\int P(x)\,dx}$ for some constant c.

Suppose y is a non-zero solution of the given equation.

Then there is an interval (a, b) such that $y(x) \neq 0$, $x \in (a, b)$.

$\therefore \quad$ For $x \in (a, b)$, we have $\dfrac{y'}{y} = -P(x)$

$\Rightarrow \qquad \qquad ln\,|y| = -\int P(x)\,dx + k$

$\Rightarrow \qquad \qquad |y| = e^{k} \cdot e^{-\int P(x)\,dx}$

$\Rightarrow \qquad \qquad y = ce^{-\int P(x)\,dx}$

where, $\qquad \qquad c = \begin{cases} e^{k} & \text{if } y > 0 \\ -e^{k} & \text{if } y < 0 \end{cases}$

Methods of Variation of Parameter to Solve the Non-homogenous Linear equation $y' + P(x)\, y = Q(x)$:

Suppose y_1 is a solution of homogeneous linear equation

$$y' + P(x)\, y = 0.$$

Let $y = uy_1$ be a solution of $y' + P(x)\, y = Q(x)$.

Then $y' = u'y_1 + uy_1'$. Substituting in the equation, we get

$$u'y_1 + uy_1' + P(x)\, uy_1 = Q(x)$$

$$\Rightarrow \quad u'y_1 + u(y_1' + P(x)\, y_1) = Q(x)$$

$$\Rightarrow \quad u'y_1 = Q(x), \ \text{ since } y_1' + P(x)\, y_1 = 0$$

$$\Rightarrow \quad u' = \frac{Q(x)}{y_1} \Rightarrow u = \int \frac{Q(x)}{y_1}\, dx + c$$

$\therefore$ The general solution of non-homogeneous linear equation is

$$y = uy_1 = e^{-\int P(x)\, dx} \int e^{\int P(x)\, dx} Q(x)\, dx + ce^{-\int P(x)\, dx}$$

Example 3.22 : *Find the general solution of* $y' + 2y = x^3\, e^{-2x}$

Solution : Here $y_1 = e^{-2x} \Rightarrow u' = x^3 \Rightarrow y = \dfrac{x^4}{4} + c$

$$\Rightarrow \quad y = ue^{-2x} = e^{-2x}\left(\frac{x^4}{4} + c\right) \text{ is a solution of the given equation.}$$

Method of integrating factor to solve linear equation :

1.   If y_1 is a solution of the homogeneous linear equation $\dfrac{dy}{dx} + P(x) = 0$, then the function $\dfrac{1}{y_1} = e^{\int P(x)\, dx}$ is called as an integrating factor for the equation $\dfrac{dy}{dx} + P(x) = Q$.

2.   Differential equation $Mdx + Ndy = 0$ is said to be exact if there exists a function $f(x, y)$ such that

$$df = \frac{\partial f}{\partial x}\, dx + \frac{\partial f}{\partial y}\, dy = Mdx + Ndy.$$

3.   Differential equation $Mdx + Ndy = 0$ is exact if and only if $\dfrac{\partial M}{\partial y} = \dfrac{\partial N}{\partial x}, \ \forall\, (x, y).$

4.　If $\dfrac{\partial M}{\partial y} \neq \dfrac{\partial N}{\partial x}$, then the equation Mdx + Ndy = 0 is non-exact.

　A function $\mu(x)$ is said to be an integrating factor (I.F.) for the equation Mdx + Ndy = 0, if $\mu(x)$ Mdx + $\mu(x)$ Ndy = 0 is an exact differential equation.

5.　If the differential equation Mdx + Ndy = 0 is exact then its general solution is given by,

$$\int M dx + \int (\text{Terms in N free from x}) \, dy = c.$$

We will discuss in detail about exact equation and integrating factor (I.F.) in Chapter 4.

Consider the first order linear differential equation

$$\frac{dy}{dx} + Py = Q \qquad \qquad \text{... (i)}$$

where, P and Q are functions of x only.

The equation (i) can be written as

$$dy + (Py - Q) \, dx = Q$$

Comparing this equation with M dx + N dy = 0, we get

　M = Py – Q　and　N = 1

$\Rightarrow$　$\dfrac{\partial M}{\partial y} = P$　　and　$\dfrac{\partial N}{\partial x} = 0$

$\Rightarrow$　$\dfrac{\partial M}{\partial y} \neq \dfrac{\partial N}{\partial x}$

$\therefore$　The given differential equation is not exact equation. But we observe that

$$\frac{\dfrac{\partial M}{\partial y} - \dfrac{\partial N}{\partial x}}{N} = \frac{P - 0}{1} = P = f(x), \text{ a function of x only}$$

$\therefore$ 　　　　I.F. $= e^{\int f(x) \, dx} = e^{\int P dx} = \dfrac{1}{y_1}$

Multiplying the equations (ii) by I.F. $e^{\int P dx}$, we have

$$e^{\int P dx} \, dy + e^{\int P dx} \, (Py - Q) \, dx = c$$

which is exact equation and its general solution is

$$\int e^{\int P dx} \, (Py - Q) \, dx + \int 0 \, dy = c$$

$$\Rightarrow \quad y \int P\, e^{\int P dx}\, dx - \int Q\, e^{\int P dx}\, dx = c$$

$$\Rightarrow \quad y \int Pe^{\int P dx}\, dx = \int Qe^{\int P dx}\, dx + c$$

Put $\qquad \int P\, dx = t$ in the first integral

$\therefore$ By definition of integration $P = \dfrac{dt}{dx} \Rightarrow Pdx = dt$

$$\Rightarrow \quad y \int e^{t}\, dt = \int Q\, e^{\int P dx}\, dx + c$$

$$\Rightarrow \quad ye^{t} = \int Q\, e^{\int P dx}\, dx + c$$

$$\Rightarrow \quad ye^{\int P dx} = \int Q\, e^{\int P dx}\, dx + c \qquad \because t = e^{\int P dx}$$

which is the general solution of linear differential equation (i), of order one.

Thus in short, "The general solution of linear differential equation"

$$\frac{dy}{dx} + Py = Q$$

where, P and Q are functions of x only is

$$ye^{\int P dx} = \int Qe^{\int P dx}\, dx + c$$

i.e. $\qquad y \cdot (I.F.) = \int Q \cdot (I.F.)\, dx + c, \quad$ where, $I.F. = e^{\int P dx}$

Note : If the linear differential equation of first order be

$$\frac{dx}{dy} + Px = Q$$

where, P and Q are functions of y only

Then $\qquad\qquad I.F. = e^{\int P dy}$

and its general solution will be

$$xe^{\int P dy} = \int Qe^{\int P dy}\, dy + c$$

i.e. $\qquad x \cdot (I.F.) = \int Q \cdot (I.F.)\, dy + c,$ where $I.F. = e^{\int P dy}$

Working Rule For Solving $\dfrac{dy}{dx} + Py = Q$:

Step - 1 : Find $I.F. = e^{\int P dx}$

Step - 2 : The general solution is

$$y \cdot (I.F.) = \int Q \cdot (I.F.)\, dx + c$$

Example 3.23 : *Solve,* $(1 + x^2)\dfrac{dy}{dx} + 2xy = \cos x.$

Solution : The given differential equation can be written as

$$\frac{dy}{dx} + \left(\frac{2x}{1 + x^2}\right) y = \frac{\cos x}{1 + x^2} \qquad \ldots \text{(i)}$$

Comparing equation (i) with $\dfrac{dy}{dx} + Py = Q$, we get

$$P = \frac{2x}{1 + x^2} \ \text{ and } \ Q = \frac{\cos x}{1 + x^2}$$

$$\therefore \qquad \text{I.F.} = e^{\int P dx} = e^{\int \frac{2x}{1 + x^2} dx} = e^{\log (1 + x^2)} = 1 + x^2$$

$$\therefore \qquad \text{I.F.} = e^{\int P dx} = 1 + x^2$$

Thus the general solution is

$$y \cdot (\text{I.F.}) = \int Q \cdot (\text{I.F.})\, dx + c$$

$$\Rightarrow \qquad y(1 + x^2) = \int \frac{\cos x}{1 + x^2} (1 + x^2)\, dx + c$$

$$\Rightarrow \qquad y(1 + x^2) = \int \cos x \, dx + c$$

$$\Rightarrow \qquad y(1 + x^2) = \sin x + c$$

which is the required solution.

3.4.2 Equation Reducible to Linear Form (Bernoulli's Equation)

Definition : A differential equation of the form

$$\frac{dy}{dx} + Py = Q y^n$$

where, P and Q are functions of x only, is called **Bernoulli's equation.**

Method of Solution of Bernoulli's Equation :

Consider the Bernoulli's equation

$$\frac{dy}{dx} + Py = Q y^n \qquad \ldots \text{(i)}$$

where, P and Q are functions of x only.

This equation can be reduced to linear differential equation of first order by some substitution.

The given equation (i) can be written as

$$y^{-n} \frac{dy}{dx} + Py^{-n+1} = Q \qquad \qquad \text{... (ii)}$$

Put $y^{-n+1} = v \Rightarrow (-n + 1) \, y^{-n} \dfrac{dy}{dx} = \dfrac{dv}{dx} \Rightarrow y^{-n} \dfrac{dy}{dx} = -\dfrac{1}{(n-1)} \dfrac{dv}{dx}$

assuming $n \neq 1$

By this substitution equation (ii) becomes

$$-\frac{1}{(n-1)} \frac{dv}{dx} + Pv = Q$$

$$\Rightarrow \qquad \frac{dv}{dx} - (n - 1) \, Pv = Q(1 - n)$$

$$\Rightarrow \qquad \frac{dv}{dx} + (1 - n) \, Pv = (1 - n) \, Q$$

which is a linear differential equation in which dependent variable is v and independent variable is x.

Hence, its I.F. is

$$\text{I.F.} = e^{\int (1 - n) \, Pdx}$$

and its general solution is

$$v \cdot (\text{I.F.}) = \int (1 - n) \, Q \cdot (\text{I.F.}) \, dx + c$$

$$\Rightarrow \qquad v e^{\int (1 - n) \, Pdx} = \int (1 - n) \, Q \cdot e^{\int (1 - n) \, Pdx} \, dx + c \quad \text{where, } v$$

$$= y^{-n+1}$$

Note : If $n = 1$, then Bernoulli's equation reduces to

$$\frac{dy}{dx} + Py = Qy$$

$$\Rightarrow \qquad \frac{dy}{dx} + (P - Q) \, y = 0$$

$$\Rightarrow \qquad \frac{dy}{y} + (P - Q) \, dx = 0$$

in which the variables are separated and hence the equation can be solved by usual method.

Example 3.24 : *Solve,* $3\dfrac{dy}{dx} + \dfrac{2y}{x+1} = \dfrac{x^3}{y^2}$

Solution : The given equation is

$$3\frac{dy}{dx} + \frac{2y}{x+1} = \frac{x^3}{y^2} \qquad \ldots \text{(i)}$$

Multiplying the given equation by y^2, we get

$$3y^2\frac{dy}{dx} + \left(\frac{2}{x+1}\right)y^3 = x^3 \qquad \ldots \text{(ii)}$$

which is a Bernoulli's equation

$$\therefore \quad \text{Put } y^3 = v \Rightarrow 3y^2\frac{dy}{dx} = \frac{dv}{dx}$$

By this substitution equation (i) reduces to

$$\frac{dv}{dx} + \left(\frac{2}{x+1}\right)v = x^3 \qquad \ldots \text{(iii)}$$

which is a linear equation of first order in which dependent variable is v and independent variable is x.

Hence, its I.F. is $\quad$ I.F. $= e^{\int \left(\frac{2}{x+1}\right)dx} = e^{2\log(x+1)} = e^{\log(x+1)^2}$

$$= (x+1)^2$$

$$\therefore \qquad\qquad \text{I.F.} = (x+1)^2$$

Hence the general solution of (iii) is

$$v(x+1)^2 = \int x^3 \cdot (x+1)^2\, dx + c$$

$$= \frac{x^6}{6} + \frac{2x^5}{5} + \frac{x^4}{4} + c$$

$$\Rightarrow \qquad v(x+1)^2 = \frac{1}{6}x^6 + \frac{2}{5}x^5 + \frac{1}{4}x^4 + c$$

$$\Rightarrow \qquad y^3(x+1)^2 = \frac{1}{6}x^6 + \frac{2}{5}x^5 + \frac{1}{4}x^4 + c \qquad \because\ v = y^3$$

which is the required general solution.

Illustrative Examples

Example 3.25 : *Solve,* $y - x\dfrac{dy}{dx} = a\left(y^2 + \dfrac{dy}{dx}\right)$

Solution : We have, $\qquad y - x\dfrac{dy}{dx} = a\left(y^2 + \dfrac{dy}{dx}\right)$

$$\Rightarrow \qquad y - ay^2 = (x + a)\frac{dy}{dx}$$

Separating the variables, we get

$$\frac{dx}{x + a} = \frac{dy}{y - ay^2}$$

Integrating both the sides, we get

$$\int \frac{dx}{x + a} = \int \frac{dy}{y - ay^2} + c_1$$

$$\Rightarrow \qquad \log(x + a) = \int \frac{dy}{y(1 - ay)} + c_1$$

$$\therefore \qquad \log(x + a) = \int \left[\frac{1}{y} + \frac{a}{1 - ay}\right] dy + c_1$$

$$(\because \text{ by partial fractions})$$

$$= \int \frac{dy}{y} + \int \frac{dy}{1 - ay} + c_1$$

$$= \log(y) - \log(1 - ay) + \log c_1$$

where, $\qquad c_1 = \log c$

$$\therefore \qquad \log(x + a) = \log(y) - \log(1 - ay) + \log c$$

$$\Rightarrow \quad \log(x + a) + \log(1 - ay) - \log(y) = \log c$$

$$\Rightarrow \qquad \log\left(\frac{(x + a)(1 - ay)}{y}\right) = \log c$$

$$\Rightarrow \qquad \frac{(x + a)(1 - ay)}{y} = c$$

$$\therefore \qquad (x + a)(1 - ay) = cy$$

which is required solution.

Example 3.26 : *Solve,* $\dfrac{dy}{dx} = e^{x-y} + x^2 e^{-y}$

Solution : We have, $\dfrac{dy}{dx} = e^{x-y} + x^2 e^{-y}$

$$= e^x \cdot e^{-y} + x^2 e^{-y}$$

$$= e^{-y}(e^x + x^2)$$

$$\therefore \qquad \frac{dy}{dx} = e^{-y}(x^2 + e^x)$$

Separating variables, we get

$$e^y \, dy = (x^2 + e^x) \, dx$$

Integrating both the sides, we get

$$\int e^y \, dy = \int (x^2 + e^x) \, dx + c$$

$$\Rightarrow \qquad e^y = \frac{x^3}{3} + e^x + c$$

which is required general solution.

Example 3.27 : *Solve,* $\log\left(\dfrac{dy}{dx}\right) = ax + by$

Solution : We have, $\qquad \log\left(\dfrac{dy}{dx}\right) = ax + by$

Taking antilog of both the sides, we get

$$\frac{dy}{dx} = e^{ax + by}$$

$$= e^{ax} \cdot e^{by}$$

$$\therefore \qquad \frac{dy}{dx} = e^{ax} \cdot e^{by}$$

Separating variables, we get

$$e^{-by} \, dy = e^{ax} \, dx$$

Integrating both the sides, we get

$$\int e^{-by} \, dy = \int e^{ax} \, dx + c$$

$$\Rightarrow \qquad \frac{e^{-by}}{-b} = \frac{e^{ax}}{a} + c$$

$$\Rightarrow \qquad b\, e^{ax} + a\, e^{-by} = k, \qquad\qquad \text{where, } k = -\,abc.$$

$$\therefore \qquad b\, e^{ax} + a\, e^{-by} = k$$

which is required solution.

Example 3.28 : *Solve,* $\cos x \cos y \, dy + \sin x \sin y \, dy = 0$

Solution : We have, $\cos x \cos y \, dy + \sin x \sin y \, dy = 0$

Separating the variables, we get

$$\frac{\cos y}{\sin y} \, dy + \frac{\sin x}{\cos x} \, dx = 0 \qquad \text{variables are separated}$$

Integrating, we get

$$\int \frac{\cos y}{\sin y} \, dy + \int \frac{\sin x}{\cos x} \, dx = \log c$$

$$\Rightarrow \qquad \log \sin y + \log \sec x = \log c$$

$$\Rightarrow \qquad \log (\sin y \sec x) = \log c$$

$$\therefore \qquad \sin y \sec x = c$$

which is required general solution.

Example 3.29 : *Solve, $(1 + e^{2x})\, dy + (1 + y^2)\, e^x\, dx = 0$.*

it being given that $y = 1$, when $x = 0$

Solution : We have, $(1 + e^{2x})\, dy + (1 + y^2)\, e^x\, dx = 0$

Separating the variables, we get

$$\frac{dy}{1 + y^2} + \frac{e^x}{1 + e^{2x}}\, dx = 0$$

Integrating both the sides, we get

$$\int \frac{dy}{1 + y^2} + \int \frac{e^x}{1 + e^{2x}}\, dx = c$$

$$\Rightarrow \qquad \tan^{-1} y + \int \frac{dt}{1 + t^2} = c \qquad \text{where, } t = e^x$$

$$\Rightarrow \qquad \tan^{-1} y + \tan^{-1} t = c$$

$$\Rightarrow \qquad \tan^{-1} y + \tan^{-1} (e^x) = c \qquad (\because\ e^x = t) \qquad \qquad \dots \text{(i)}$$

which is the general solution of the given differentiation equation.

We have given that $y = 1$, when $x = 0$.

$\therefore$ Putting $x = 0$ and $y = 1$ in equation (i), we get

$$\tan^{-1} (1) + \tan^{-1} (e^0) = c$$

$$\Rightarrow \qquad \tan^{-1} (1) + \tan^{-1} (1) = c$$

$$\Rightarrow \qquad \frac{\pi}{4} + \frac{\pi}{4} = c$$

$$\Rightarrow \qquad \frac{\pi}{2} = c$$

$$\therefore \qquad c = \frac{\pi}{2}$$

Substituting this value of c in equation (i), we get

$$\tan^{-1} y + \tan^{-1} (e^x) = \frac{\pi}{2}$$

which is the required solution.

Example 3.30 : *Solve,* $x\,dy - y\,dx = \sqrt{x^2 + y^2}\,dx$

Solution : The given differential equation can be written as

$$\frac{dy}{dx} = \frac{y + \sqrt{x^2 + y^2}}{x} \qquad \dots (i)$$

Clearly equation (i) is a homogeneous differential equation as $y + \sqrt{x^2 + y^2}$ and x are homogeneous functions of same degree 1.

$\therefore$ Putting $y = vx$ so that $\dfrac{dy}{dx} = v + x\dfrac{dv}{dx}$ in equation (i), we get

$$v + x\frac{dv}{dx} = \frac{vx + \sqrt{x^2 + v^2 x^2}}{x}$$

$$= v + \sqrt{1 + v^2}$$

$$\Rightarrow \qquad x\frac{dv}{dx} = \sqrt{1 + v^2}$$

Separating the variables, we get

$$\frac{dv}{\sqrt{1 + v^2}} = \frac{dx}{x}$$

Integrating both the sides, we get

$$\int \frac{dv}{\sqrt{1 + v^2}} = \int \frac{dx}{x} + \log c$$

$$\Rightarrow \qquad \log\left(v + \sqrt{1 + v^2}\right) = \log x + \log c$$

$$\Rightarrow \qquad \log\left(\frac{y}{x} + \sqrt{1 + \frac{y^2}{x^2}}\right) = \log x + \log c$$

$$\Rightarrow \qquad \log\left(\frac{y}{x} + \frac{\sqrt{x^2 + y^2}}{x}\right) = \log x + \log c$$

$$\Rightarrow \qquad \log\left(\frac{y + \sqrt{x^2 + y^2}}{x}\right) = \log x + \log c$$

$$\Rightarrow \quad \log\left(y + \sqrt{x^2 + y^2}\right) - \log x - \log x = \log c$$

$$\Rightarrow \qquad \log\left(\frac{y + \sqrt{x^2 + y^2}}{x^2}\right) = \log c$$

$$\Rightarrow \qquad \frac{y + \sqrt{x^2 + y^2}}{x^2} = c$$

$$\Rightarrow \qquad y + \sqrt{x^2 + y^2} = cx^2$$

which is the required solution.

Example 3.31 : *Solve,* $x \dfrac{dy}{dx} = y - x \, tan \left(\dfrac{y}{x} \right)$

Solution : The given differential equation can be written as

$$\frac{dy}{dx} = \frac{y - x \, tan \left(\dfrac{y}{x} \right)}{x} \qquad \text{... (i)}$$

Clearly equation (i) is a homogeneous differential equation as $y - x \, tan \left(\dfrac{y}{x} \right)$ and x are homogeneous functions of same degree 1.

$\therefore$ Putting $y = vx$ so that $\dfrac{dy}{dx} = v + x \dfrac{dv}{dx}$ in equation (i), we get

$$v + x \frac{dv}{dx} = \frac{vx - x \, tan \left(\dfrac{vx}{x} \right)}{x}$$

$$= v - tan \, v$$

$$\Rightarrow \qquad x \frac{dv}{dx} = - tan \, v$$

Separating the variables, we get

$$\frac{dv}{tan \, v} + \frac{dx}{x} = 0$$

$$\Rightarrow \qquad \frac{cos \, v}{sin \, v} \, dv + \frac{dx}{x} = 0$$

Integrating both the sides, we get

$$\int \frac{cos \, v}{sin \, v} \, dv + \int \frac{dx}{x} = log \, c$$

$$\Rightarrow \qquad log \, (sin \, v) + log \, (x) = log \, c$$

$$\Rightarrow \qquad log \, (x \, sin \, v) = log \, c$$

$$\Rightarrow \qquad x \, sin \, v = c$$

$$\Rightarrow \qquad x \, sin \left(\frac{y}{x} \right) = c$$

which is required solution.

Example 3.32 : *Solve,* $\left(x \sin\left(\dfrac{y}{x}\right) \right) dy = \left(y \sin\left(\dfrac{y}{x}\right) - x \right) dx$

Solution : The given differential equation can be written as

$$\frac{dy}{dx} = \frac{y \sin\left(\dfrac{y}{x}\right) - x}{x \sin\left(\dfrac{y}{x}\right)} \qquad \ldots \text{(i)}$$

Clearly equation (i) is homogeneous differential equation as $y \sin\left(\dfrac{y}{x}\right) - x$ and $x \sin\left(\dfrac{y}{x}\right)$ are homogeneous functions of same degree 1.

$\therefore$ Putting $y = vx$ so that $\dfrac{dy}{dx} = v + x\dfrac{dv}{dx}$ in equation (i), we get

$$v + x\frac{dv}{dx} = \frac{vx \sin v - x}{x \sin v}$$

$$\Rightarrow \qquad v + x\frac{dv}{dx} = \frac{v \sin v - 1}{\sin v}$$

$$\Rightarrow \qquad x\frac{dv}{dx} = \frac{v \sin v - 1}{\sin v} - v$$

$$= -\frac{1}{\sin v}$$

Separating the variables, we get

$$\sin v \, dv + \frac{dx}{x} = 0$$

Integrating both the sides, we get

$$\int \sin v \, dv + \int \frac{dx}{x} = c$$

$$\Rightarrow \qquad -\cos v + \log x = c$$

$$\Rightarrow \qquad \cos v = \log x - c$$

$$\Rightarrow \qquad \cos v = \log x + c_1$$

$$\therefore \qquad \cos\left(\frac{y}{x}\right) = \log x + c_1$$

which is the required solution.

Example 3.33 : *Solve,* $(2x - y)\, e^{y/x}\, dx + (y + x\, e^{y/x})\, dy = 0$

Solution : The given differential equation can be written as

$$\frac{dy}{dx} = -\frac{(2x - y)\, e^{y/x}}{y + x\, e^{y/x}} \qquad\qquad \ldots \text{(i)}$$

Clearly equation (i) is a homogeneous differential equation as $(2x - y)\, e^{y/x}$ and $y + x\, e^{y/x}$ are homogeneous functions of same degree 1.

$\therefore$ Putting $y = vx$ so that $\dfrac{dy}{dx} = v + x\,\dfrac{dv}{dx}$ in equation (i), we get

$$v + x\frac{dv}{dx} = -\frac{[2x - vx]\, e^{vx/x}}{vx + x\, e^{vx/x}}$$

$$= -\frac{(2 - v)\, e^{v}}{v + e^{v}}$$

$$\Rightarrow \qquad x\frac{dv}{dx} = -\frac{(2 - v)\, e^{v}}{v + e^{v}} - v$$

$$= -\left[\frac{(2 - v)\, e^{v} + v(v + e^{v})}{v + e^{v}}\right]$$

$$= -\frac{2e^{v} + v^2}{v + e^{v}}$$

$$\Rightarrow \qquad x\frac{dv}{dx} = -\frac{2e^{v} + v^2}{v + e^{v}}$$

Separating the variables, we get

$$\frac{v + e^{v}}{v^2 + 2e^{v}}\, dv = -\frac{dx}{x}$$

$$\Rightarrow \qquad \frac{v + e^{v}}{v^2 + 2e^{v}}\, dv + \frac{dx}{x} = 0$$

Integrating both the sides, we get

$$\int \frac{v + e^{v}}{v^2 + 2e^{v}}\, dv + \int \frac{dx}{x} = c_1$$

$$\Rightarrow \qquad \frac{1}{2}\int \frac{2v + 2e^{v}}{v^2 + 2e^{v}}\, dv + \int \frac{dx}{x} = c_1$$

$$\Rightarrow \qquad \frac{1}{2}\log (v^2 + 2e^{v}) + \log (x) = c_1$$

$$\Rightarrow \qquad \log (v^2 + 2e^{v}) + 2 \log (x) = 2c_1$$

$$\Rightarrow \qquad \log x^2(v^2 + 2e^v) = \log c \qquad \text{where, } 2c_1 = \log c$$

$$\Rightarrow \qquad x^2(v^2 + 2e^v) = c$$

$$\Rightarrow \qquad y^2 + 2x^2\, e^{y/x} = c$$

which is the required solution.

Example 3.34 : *Solve,* $\cosec x \dfrac{dy}{dx} = y + \cos x$

Solution : The given differential equation can be written as

$$\dfrac{dy}{dx} - (\sin x)\, y = \sin x \cos x \qquad \ldots (i)$$

which is a linear differential equation of first order.

Comparing equation (i) with $\dfrac{dy}{dx} + Py = Q$, we get

$$P = -\sin x \ \text{ and } \ Q = \sin x \cos x$$

$$\therefore \qquad \text{I.F.} = e^{\int Pdx} = e^{\int -\sin x\, dx} = e^{\cos x}$$

$$\therefore \qquad \text{I.F.} = e^{\cos x}$$

Hence the required general solution is

$$y \cdot (\text{I.F.}) = \int Q \cdot (\text{I.F.})\, dx + c$$

$$\Rightarrow \qquad y\, e^{\cos x} = \int \sin x \cos x\, e^{\cos x}\, dx + c$$

Put $\cos x = t \Rightarrow -\sin x\, dx = dt$

$$= \int -t\, e^t\, dt + c$$

$$\text{Integrating by parts} = -[t\, e^t - e^t] + c$$

$$= -(t - 1)\, e^t + c$$

$$= (1 - t)\, e^t + c$$

$$= (1 - \cos x)\, e^{\cos x} + c \qquad \because \cos x = t$$

$$\Rightarrow \qquad y\, e^{\cos x} = (1 - \cos x)\, e^{\cos x} + c$$

$$\Rightarrow \qquad y = 1 - \cos x + c e^{-\cos x}$$

Example 3.35 : *Solve,* $dx + x\, dy = e^{-y} \sec^2 y\, dy.$

Solution : The given differential equation can be written as

$$\dfrac{dx}{dy} + x = e^{-y} \sec^2 y \qquad \ldots (i)$$

which is a linear differential equation in which dependent variable is x and independent variable is y.

Comparing the equation (i) $\dfrac{dx}{dy} + Px = Q$, we get

$$P = 1 \ \text{ and } \ Q = e^{-y} \sec^2 y$$

$\therefore$ 　　　　I.F. $= e^{\int Pdy} = e^{\int 1 \, dy} = e^{y}$

$\therefore$ 　　　　I.F. $= e^{y}$

Hence the required general solution is

$$x \cdot (\text{I.F.}) = \int Q \cdot (\text{I.F.}) \, dy + c$$

$\Rightarrow$ 　　　$xe^{y} = \int (e^{-y} \sec^2 y) \, e^{y} \, dy + c$

$$= \int \sec^2 y \, dy + c$$

$$= \tan y + c$$

$\Rightarrow$ 　　　$xe^{y} - \tan y = c$

which is required general solution.

Example 3.36 : *Solve, $2(1 + x)\dfrac{dy}{dx} - (1 + 2x)y = x^2 \sqrt{(1 + x)}$*

Solution : The given differential equation can be written as

$$\frac{dy}{dx} - \left[\frac{(1 + 2x)}{2(1 + x)}\right] y = \frac{x^2}{2\sqrt{1 + x}} \qquad \qquad \dots \text{(i)}$$

which is a linear differential equation in which dependent variable is y and independent variable is x.

Comparing the given equation (i) with $\dfrac{dy}{dx} + Py = Q$, we get

$$P = -\frac{(1 + 2x)}{2(1 + x)} \ \text{ and } \ Q = \frac{x^2}{2\sqrt{1 + x}}$$

$\therefore$ 　　　$\text{I.F.} = e^{\int Pdx} = e^{-\int \frac{1 + 2x}{2(1 + x)} \, dx}$

$$= e^{-\int \frac{2 + 2x - 1}{2(1 + x)} \, dx}$$

$$= e^{-\int \left[1 - \frac{1}{2(1 + x)}\right] dx}$$

$$= e^{-\left[x - \frac{1}{2} \log(1 + x)\right]} = e^{-x} \cdot e^{\frac{1}{2} \log(1 + x)}$$

$$= e^{-x} \, e^{\log \sqrt{1 + x}}$$

$$= e^{-x} \sqrt{1 + x}$$

$$\therefore \qquad \text{I.F.} = e^{\int P dx} = e^{-x} \sqrt{1 + x}$$

Hence the required general solution is

$$y \cdot (\text{I.F.}) = \int Q \cdot (\text{I.F.}) \, dx + c$$

$$\Rightarrow \qquad y\sqrt{1 + x} \, e^{-x} = \int \frac{x^2}{2\sqrt{1 + x}} \cdot \sqrt{1 + x} \, e^{-x} \, dx + c$$

$$= \frac{1}{2} \int x^2 \, e^{-x} \, dx + c$$

Integrating by parts $= \dfrac{1}{2} \left[x^2 (-e^{-x}) - \int 2x \, (-e^{-x}) \, dx \right] + c$

$$= -\frac{1}{2} x^2 \, e^{-x} + \int x \, e^{-x} \, dx + c$$

$$= -\frac{1}{2} x^2 \, e^{-x} + \left[-x \, e^{-x} + \int e^{-x} \, dx \right] + c$$

$$= -\frac{1}{2} x^2 \, e^{-x} - x \, e^{-x} - e^{-x} + c$$

$$= -\frac{1}{2} (x^2 + 2x + 2) \, e^{-x} + c$$

$$\Rightarrow \qquad y\sqrt{1 + x} \, e^{-x} = -\frac{1}{2} (x^2 + 2x + 2) \, e^{-x} + c$$

$$\Rightarrow \qquad y\sqrt{1 + x} = -\frac{1}{2} (x^2 + 2x + 2) + ce^{x}$$

which is the required general solution.

Example 3.37 : *Solve,* $(x + y + 1)\dfrac{dy}{dx} = 1$

Solution : The given differential equation can be written as,

$$\frac{dx}{dy} = x + y + 1$$

i.e. $\qquad \dfrac{dx}{dy} - x = y + 1 \qquad\qquad\qquad\qquad \dots \text{(i)}$

which is a linear differential equation of first order in which dependent variable is x and independent variable is y.

Comparing equation (i) with $\dfrac{dx}{dy} + Px = Q$, we get

$$P = -1, \text{ and } Q = y + 1$$

$$\therefore \qquad \text{I.F.} = e^{\int Pdy} = e^{\int(-1)\,dy} = e^{-y}$$

$$\therefore \qquad \text{I.F.} = e^{\int Pdy} = e^{-y}$$

Hence required general solution is

$$x(\text{I.F.}) = \int Q(\text{I.F.})\,dy + c$$

$$\Rightarrow \qquad xe^{-y} = \int (y + 1)\,e^{-y}\,dy + c$$

Integrating by parts

$$\Rightarrow \qquad x\,e^{-y} = -(y+1)\,e^{-y} - \int -e^{-y}\,dy + c$$

$$= -(y+1)\,e^{-y} - e^{-y} + c$$

$$= -(y+2)\,e^{-y} + c$$

$$\Rightarrow \qquad x = -(y+2) + ce^{y}$$

which is required solution.

Example 3.38 : *Solve,* $y + 2\dfrac{dy}{dx} = y^3\,(x - 1)$

Solution : The given differential equation is

$$y + 2\frac{dy}{dx} = y^3\,(x - 1) \qquad\qquad \dots \text{(i)}$$

Dividing the equation (i) by y^3, we get

$$2y^{-3}\frac{dy}{dx} + y^{-2} = x - 1 \qquad\qquad \dots \text{(ii)}$$

which is a Bernoulli's equation.

$$\therefore \quad \text{Put } y^{-2} = v \Rightarrow -2y^{-3}\frac{dy}{dx} = \frac{dv}{dx}$$

By this substitution equation (ii) becomes

$$-\frac{dv}{dx} + v = x - 1$$

$$\Rightarrow \qquad \frac{dv}{dx} - v = 1 - x \qquad\qquad \dots \text{(iii)}$$

which is a linear equation in which dependent variable is v and independent variable is x. Hence its I.F. is

$$\text{I.F.} = e^{\int -1 \cdot dx} = e^{-x}$$

Hence the general solution of (iii) is

$$ve^{-x} = \int (1 - x)\,e^{-x}\,dx + c$$

By integrating by parts

$$\Rightarrow \qquad ve^{-x} = -(1-x)e^{-x} - \int e^{-x}\, dx + c$$

$$\Rightarrow \qquad ve^{-x} = -(1-x)e^{-x} + e^{-x} + c$$

$$= (x-1)e^{-x} + e^{-x} + c$$

$$= xe^{-x} + c$$

$$\Rightarrow \qquad ve^{-x} = xe^{-x} + c$$

$$\Rightarrow \qquad y^{-2}e^{-x} = xe^{-x} + c \qquad\qquad \because\ y^{-2} = v$$

$$\Rightarrow \qquad y^{-2} = x + ce^{x}$$

$$\Rightarrow \qquad y^{2}(x + ce^{x}) = 1$$

which is required general solution.

Example 3.39 : *Solve,* $(x^3y^3 + xy)\dfrac{dy}{dx} = 1$

Solution : The given equation can be written as

$$\frac{dx}{dy} = x^3y^3 + xy$$

$$\Rightarrow \qquad \frac{dx}{dy} - yx = x^3y^3 \qquad\qquad \ldots\text{(i)}$$

which is a Bernoulli's equation in which the dependent variable being x and independent variable being y.

Dividing equation (i) by x^3, we get

$$x^{-3}\frac{dx}{dy} - y\,x^{-2} = y^3 \qquad\qquad \ldots\text{(ii)}$$

Put $x^{-2} = v$

$$\Rightarrow \qquad -2x^{-3}\frac{dx}{dy} = \frac{dv}{dy}$$

$$\Rightarrow \qquad x^{-3}\frac{dx}{dy} = -\frac{1}{2}\frac{dv}{dy}$$

By this substitution equation (ii) becomes

$$-\frac{1}{2}\frac{dv}{dy} - yv = y^3$$

$$\Rightarrow \qquad \frac{dv}{dy} + 2yv = -2y^3 \qquad\qquad \ldots\text{(iii)}$$

which is a linear equation in which dependent variable is v and independent variable is y. Hence, its I.F. is

$$\text{I.F.} = e^{\int 2y\, dy} = e^{y^2}$$

Hence the general solution of (iii) is

$$v \,(\text{I.F.}) = \int (-2y^3) \,(\text{I.F.}) \, dy + c$$

$$\Rightarrow \qquad v(e^{y^2}) = \int (-2y^3) \, e^{y^2} \, dy + c$$

$$\Rightarrow \qquad v \, e^{y^2} = -2 \int y^3 \, e^{y^2} \, dy + c$$

$$= -2 \int t \, e^t \cdot \frac{dt}{2} + c \quad \because \; y^2 = t \Rightarrow 2y \, dy = dt$$

$$= -\int t \, e^t \, dt + c$$

$$= -[t \, e^t - \int e^t \, dt] + c$$

$$= -[t \, e^t - e^t] + c = -(t-1) \, e^t + c$$

$$= -(y^2 - 1) \, e^{y^2} + c$$

$$\Rightarrow \qquad v \, e^{y^2} = -(y^2 - 1) \, e^{y^2} + c$$

$$\Rightarrow \qquad x^{-2} \, e^{y^2} = -(y^2 - 1) \, e^{y^2} + c \qquad\qquad \because \; v = x^{-2}$$

$$\Rightarrow \qquad x^{-2} = (1 - y^2) + c \, e^{-y^2}$$

$$\Rightarrow \; x^2 \, [(1 - y^2) + c \, e^{-y^2}] = 1$$

which is required general solution.

3.5 Existence and Uniqueness of Solutions of Non-linear Equations

Given a differential equation $y' = f(x, y)$, it is not always possible to obtain its solution which is defined for all $x \in \mathbb{R}$. There are certain conditions which ensures the existence of solution of the differential equation in the prescribed interval.

Theorem 5 (i) If $f(x, y)$ is continuous on an open rectangle $R : \{(x, y) \mid a < x < b, c < y < d\}$ and $(x_0, y_0) \in R$, then there exists a solution $y = y(x)$ of the differential equation $y' = f(x, y)$ such that $y(x_0) = y_0$.

(ii) Further, if $f(x, y)$ and $\dfrac{\partial f}{\partial y}$ are both continuous on R, then the differential equation $y' = f(x, y)$ has a unique solution $y = y(x)$, $x \in (a, b)$ such that $y(x_0) = y_0$.

Remark : Part (i) of the theorem guarantees the existence of the solution of $y' = f(x, y)$ while part (ii) guarantees the uniqueness of the solution that satisfies the initial condition $y(x_0) = y_0$. Hence, the theorem is called as existence and uniqueness theorem.

Example : Consider the initial value problem

$$y' = \frac{x^2 - y^2}{1 + x^2 + y^2}, \; y(x_0) = y_0$$

Here, $f(x, y) = \frac{x^2 - y^2}{1 + x^2 + y^2}$. Since, $1 + x^2 + y^2 \neq 0$, the function $f(x, y)$ is continuous everywhere in $\mathbb{R}^2$. Moreover $\dfrac{\partial f}{\partial y} = \dfrac{-2y(1 + 2x^2)}{(1 + x^2 + y^2)^2}$ is also continuous everywhere on $\mathbb{R}^2$. Hence, there is unique solution $y = y(x)$ of the differential equation passing through any point $(x_0, y_0) \in \mathbb{R}^2, x \in \mathbb{R}$.

Example : Consider the equation $y' = \dfrac{x^2 - y^2}{x^2 + y^2}, \; y(x_0) = y_0$.

Here $f(x, y) = \dfrac{x^2 - y^2}{x^2 + y^2}$ and $\dfrac{\partial f}{\partial y}(x, y) = \dfrac{-4x^2 y}{(x^2 + y^2)^2}$ are continuous everywhere except at $(0, 0)$. Thus, if R is an open rectangle that does not contain $(0, 0)$ and $(x_0, y_0) \in R$ is any point, then there is a unique solution $y = y(x)$ of given differential equation such that $y(x_0) = y_0$.

Example : Consider the initial value problem

$$y' = \frac{x + y}{x - y}, \; y(x_0) = y_0$$

Here $f(x, y) = \dfrac{x + y}{x - y}, \; \dfrac{\partial f}{\partial y}(x, y) = \dfrac{2x}{(x - y)^2}$ are continuous everywhere except on the line $y = x$. If $y_0 \neq x_0$, then there is an open rectangle containing (x_0, y_0) that does not intersect the line $y = x$.

There is unique solution on some open interval containing x_0.

Illustrative Examples

Example 3.40 : *Consider the differential equation $y' = 2xy^2$. Show that $y = 0 \; \forall \, x \in \mathbb{R}$ is the only solution of given differential equation that passes through $(0, 0)$.*

Solution : Here $f(x, y) = 2xy^2$ and $\dfrac{\partial f}{\partial y} = 4xy$ are continuous everywhere on $\mathbb{R}^2$. Therefore, there is unique solution passing through any point in $\mathbb{R}^2$. Note here, that $y \equiv 0$ is a solution of the given differential equation and it satisfies the condition $y(0) = 0$. By existence and uniqueness theorem there is no other solution passing through $(0, 0)$.

Example 3.41 : *Consider the initial value problem.*

$$y' = \frac{10}{3} xy^{2/5}, \quad y(x_0) = y_0.$$

(a) For what values of (x_0, y_0) does the problem has a solution ?

(b) For what values of (x_0, y_0) does the problem has a unique solution on some open interval containing x_0 ?

Solution : (a) Since, $f(x, y) = \dfrac{10}{3} xy^{2/3}$ is continuous for all (x, y), by existence theorem, there is a solution passing through any point (x_0, y_0).

(b) Further $\dfrac{\partial f}{\partial y} = \dfrac{4}{3} xy^{-3/5}$ is not continuous at (x, y) if $y = 0$. Then there is unique solution passing through any point (x_0, y_0) with $y_0 \neq 0$.

Think Over It

1. Sketch the solution curves of all the differential equations that you studied in this chapter.

2. Let $f(x) = \sin x$, $g(x) = \cos x$, $h(x) = e^x$. Then using differential equations, prove that :

   (a) $f'(x) = -g(x)$ (b) $g'(x) = -f(x)$

   (c) $f''(x) + f(x) = 0$ (d) $g''(x) + g(x) = 0$

   (e) $h(x + y) = h(x) \cdot h(y)$ (f) $h'(x) = h(x)$

Summary

1. General solution of the equation $g(y)\, y' = f(x)$ is
$$\int g(y)\, dy = \int f(x)\, dx + c.$$

2. General solution of the linear homogeneous equation
$$y' + p(x)\, y = 0 \text{ is } y = ce^{-\int p(x)\, dx}.$$

3. General solution of non-homogeneous linear equation
$$y' + p(x)\, y = \phi(x) \text{ is } y(x) = e^{-\int p(x)\, dx} \int e^{\int p(x)\, dx}\, \phi(x)\, dx + ce^{-\int p(x)\, dx}.$$

4. If $f(x, y)$ and $\dfrac{\partial f}{\partial y}$ are continuous in an open rectangle

   $R : \{(x, y) \mid a < x < b, c < y < d\}$, then for any $(x_0, y_0) \in R$ there exists unique solution of the initial value problem $y' = f(x, y)$, $y(x_0) = y_0$.

Exercise

[A] Say True or False : Justify !

1. Every differential equation has a solution passing through given point (x_0, y_0).

2. If solution of the differential equation exists then it is unique.

3. Degree of the differential equation $y'' + (y')^2 + y = 0$ is 2.

4. $y\, dx + x\, dy = 0$ has unique solution passing through the point $(1, 1)$.

5. Slope of the tangent to the solution curve of the differential equation $M dx + N dy = 0$ is $-\dfrac{M}{N}$.

[B] Multiple Choice Questions : Choose the Correct Alternative

1. The function $f(x, y) = e^{x/y}$ is homogeneous function of degree ......
   - (a) 1
   - (b) 0
   - (c) 2
   - (d) none of these

2. The function $f(x, y)$ has how many second order partial derivatives ?
   - (a) 2
   - (b) 4
   - (c) 3
   - (d) none of these

3. The ordinary differential equation always involves ......
    (a) one independent variable.
    (b) two independent variables.
    (c) more than two independent variables.
    (d) none of these.

4. If the differential equation is of the form $\left(\dfrac{d^n y}{dx^n}\right)^m = c$, then its
    ......
    (a) order n but degree not m   (b) degree m but order not n
    (c) order n and degree m   (d) none of these

5. The general solution of first order differential equations contains
    (a) one arbitrary constant   (b) two arbitrary constants
    (c) three arbitrary constants   (d) none of these

6. The equation $(x^2 + y^2)\dfrac{dy}{dx} = xy$ is ......
    (a) homogeneous   (b) linear
    (c) reducible to linear   (d) none of these

7. Consider differential equation $\dfrac{dy}{dx} = \dfrac{y}{x}$. Then ......
    (a) it has unique solution passing through $(0, 0)$.
    (b) it has more than one solution passing through $(1, 1)$.
    (c) it has unique solution through $(2, 1)$.
    (d) none of the above.

8. Differential equation for the family of all straight lines passing
    through the point $(1, 1)$ is ......
    (a) $(y - 1)\, dx = (x - 1)\, dy$ (b) $\dfrac{dy}{dx} = \dfrac{x - 1}{y - 1}$

    (c) $y = 1 + x\,\dfrac{dy}{dx}$ (d) $y = 1 + (x + 1)\,\dfrac{dy}{dx}$

[C] Theory Questions :

1. Define differential equation.

2. Find the differential equation for the family of all circles with
    center at the origin.

3. Find the differential equation for the family of all straight
    lines passing through the point $(3, 2)$.

4. Suppose that $f(x)$, $g(y)$ are continuous on intervals (a, b) and (c, d) respectively. Let $F(x) = \int f(x)\, dx$ and $G(y) = \int g(y)\, dy$, let $y_0 \in (c, d)$ such that $g(y_0) \neq 0$ and $c = G(y_0) - F(x_0)$. Then prove that, there is a function $y = y(x)$ defined on some interval $(a_1, b_1) = (a, b)$ such that $y(x_0) = y_0$ and $G(y) = F(x) + c$, $x \in (a_1, b_1)$

5. State the existence and uniqueness theorem for the first order ordinary differential equation $y' = f(x, y)$.

6. Define : (i) Homogeneous function of two variables.

   (ii) Homogeneous function of three variables.

   Give suitable illustrations.

7. Define : Partial derivatives $\dfrac{\partial f}{\partial x}$ and $\dfrac{\partial f}{\partial y}$ of the function $u = f(x, y)$.

8. What is an ordinary differential equation ?

9. Define : (i) The order and (ii) The degree of a differential equation.

10. Explain the method of solving the equation $M\, dx + N\, dy = 0$, when the variables in it are separable.

11. Define a homogeneous differential equation of first order and first degree and explain the method of solving it.

12. Explain the method of solving the equation $M\, dx + N\, dy = 0$, when M and N are homogeneous functions of the same degree, in x and y.

13. Explain the method of solving the equation
$$\frac{dy}{dx} = \frac{a_1 x + b_1 y + c_1}{a_2 x + b_2 y + c_2}$$

   when $a_1 b_2 - a_2 b_1 \neq 0$.

14. Reduce the equation $\dfrac{dy}{dx} = \dfrac{a_1 x + b_1 y + c_1}{a_2 x + b_2 y + c_2}$

   to homogeneous differential equation.

15. Define a linear differential equation. Explain the method to solve the linear differential equation.
$$\frac{dy}{dx} + Py = Q$$

   where, P and Q are functions of x alone.

16. Find an integrating factor of the linear equation

$$\frac{dy}{dx} + Py = Q$$

P and Q being functions of x alone. Hence, solve the equation.

17. Define Bernoulli's equation; show that it can be reduced to linear equation. Hence solve Bernoulli's equation.

[D] Numerical Problems :

1. Which of the following functions are homogeneous functions. If they are homogeneous then determine the degree of it :

(i) $f(x, y) = 5x^4 - 7x^3y + 3xy^3$

(ii) $f(x, y) = (5x^3 + 5xy^2)^{-4}$

(iii) $f(x, y) = \sin\left(\dfrac{x^2 + y^2}{x^2 - 3y^2}\right)$

(iv) $u = \dfrac{\sqrt{x^{1/2} + y^{1/2}}}{\sqrt{x^{1/3} + y^{1/3}}}$

(v) $f(x, y, z) = x^3 + y^3 + z^3 - 3xyz$

(vi) $f(x, y, z) = 3x^2yz + 5xy^2z + 4z^4$.

2. Find $\dfrac{\partial u}{\partial x}$ and $\dfrac{\partial u}{\partial y}$, if

(i) $u = \tan^{-1}\left(\dfrac{x^2 + y^2}{x + y}\right)$

(ii) $u = \dfrac{1 - xy}{x + y}$

(iii) $u = \log(x^2 + y^2)$

3. Find $\dfrac{\partial^2 u}{\partial x\, \partial y}$, when

(i) $u = \tan^{-1}\left(\dfrac{xy}{\sqrt{1 + x^2 + y^2}}\right)$

(ii) $u = \log\dfrac{(1 + x^2)(1 + y^2)}{xy}$

4. When $u = x^y + y^x$. Verify that $\dfrac{\partial^2 u}{\partial x\, \partial y} = \dfrac{\partial^2 u}{\partial y\, \partial x}$

5. Determine order and degree of each of the following differential equations :

(i) $\left(\dfrac{dy}{dx}\right)^2 + 5y = \sin x$

(ii) $\left(\dfrac{d^2y}{dx^2}\right)^3 + 2\left(\dfrac{dy}{dx}\right)^4 + y = \sin x$

(iii) $(x^3 - y^3)\, dx + (xy^2 - x^2y)\, dy = 0$

(iv) $y = px \pm c\sqrt{p^2 + 1},$ where, $p = \dfrac{dy}{dx}$

(v) $\dfrac{d^2y}{dx^2} = \sqrt{1 + \left(\dfrac{dy}{dx}\right)^2}$

(vi) $\dfrac{d^2y}{dx^2} + \cos\left(\dfrac{dy}{dx}\right) = 0$

6. Find the differential equation of the family of all straight lines making equal intercepts on the co-ordinate axes.

7. Find the differential equation that will represent the family of all circles having their centres on the x-axis and radii equal to unity.

8. Find the differential equation of the family of curves
$$y = ae^x + be^{-x}$$
where, a and b are arbitrary constants.

9. Find the differential equation of the family of curves
$$y = e^x (a \cos x + b \sin x)$$
where, a and b are arbitrary constants.

10. Solve the following differential equations :

(i) $\dfrac{dy}{dx} = \log (x + 1)$

(ii) $(x - 1)\dfrac{dy}{dx} = 2x^3y$

(iii) $(x^2 - yx^2)\, dy + (y^2 + xy^2)\, dx = 0$

(iv) $\sqrt{a + x}\,\dfrac{dy}{dx} + x = 0$

(v) $\cos x\, (1 + \cos y)\, dx - \sin y\, (1 + \sin x)\, dy = 0$

(vi) $\dfrac{dy}{dx} = \dfrac{e^x (\sin^2 x + \sin 2x)}{y(2 \log y + 1)}$

(vii) $(1 + y^2)\, (1 + \log x)\, dx + x\, dy = 0,$ it being given that $y = 1$, when $x = 0$.

(viii) $\dfrac{dy}{dx} = y \sin 2x$, it being given that $y(0) = 1$

(ix) $(1 + x^2) \sec^2 y \, dy + 2x \tan y \, dx = 0$, given that $y = \dfrac{\pi}{4}$, when $x = 1$.

(x) $(x - y)^2 \dfrac{dy}{dx} = a^2$

(xi) $\cos (x + y) \, dy = dx$

(xii) $\dfrac{dy}{dx} = \sin (x + y) + \cos (x + y)$

11. Solve the following differential equations :

(i) $(x^2 + y^2) \dfrac{dy}{dx} = xy$

(ii) $(y^2 - x^2) \, dy = 3xy \, dx$

(iii) $(x^3 + y^3) \, dy - x^2 y \, dx = 0$

(iv) $(3xy + y^2) \, dx + (x^2 + xy) \, dy = 0$

(v) $x^2 dy + y(x + y) \, dx = 0$

(vi) $y^2 dx + (x^2 + xy + y^2) \, dy = 0$

(vii) $\left(x\sqrt{x^2 + y^2} - y^2\right) dx + xy \, dy = 0$

(viii) $(x^3 - 3xy^2) \, dx = (y^3 - 3x^2 y) \, dy$

12. Solve the following differential equations :

(i) $(2x + y - 3) \, dy = (x + 2y - 3) \, dx$

(ii) $\dfrac{dy}{dx} = \dfrac{2x + 2y - 2}{3x + y - 5}$

(iii) $\dfrac{dy}{dx} = \dfrac{x - y + 3}{2x - 2y + 5}$

(iv) $(x - y) \, dy = (x + y + 1) \, dx$

(v) $(6x + 2y - 10) \dfrac{dy}{dx} = 2x - 9y + 20$

13. Solve the following differential equations :

(i) $(4x + 3y + 1) \, dx + (3x + 2y + 1) \, dy = 0$

(ii) $\dfrac{dy}{dx} = \dfrac{2x - y}{x + 2y - 5}$

(iii) $(1 + 4xy + 2y^2) \, dx + (1 + 4xy + 2x^2) \, dy = 0$

    (iv) $[1 + e^{x/y}]\, dx + e^{x/y}\left[1 - \left(\dfrac{x}{y}\right)\right] dy = 0$

    (v) $(y \sin 2x)\, dx - (1 + y^2 + \cos^2 x)\, dy = 0$

14. Find all solutions of $y' = 2xy^2$

15. Find all solutions of $y' = \dfrac{1}{2}\, x(1 - y^2).$

16. Find all solutions of :

    (i) $(3y^3 + 3y \cos y + 1)\, y' = \dfrac{(2x + 1)\, y}{1 + x^2} = 0$

    (ii) $x^2 yy' = (y^2 - 1)^{3/2}$

    (iii) $y' \, ln\, |y| + x^2 y = 0$

17. Solve the following initial value problems :

    (i) $y' = \dfrac{x^2 + 2x + 3}{y - 2}\, , \, y(1) = 4$

    (ii) $y' + x(y^2 + y) = 0,\, y(2) = 1$

18. Solve the initial value problems and find the interval of the validity of the solution :

    (i) $y'(x^2 + 2) + 4x(y^2 + 2y + 1) = 0,\;\; y(1) = -1$

    (ii) $y' = -2x(y^2 - 3y + 2),\, y(0) = 3$

    (iii) $x + yy' = 0,\;\; y(3) = -4$

    (iv) $(x + y)\, (x - 2)\, y' + y = 0,\;\; y(1) = -3$

19. Solve the following differential equations explicitly :

    (i) $y' = \dfrac{(1 + y^2)}{(1 + x^2)}$ (ii) $y' \sqrt{1 - x^2} + \sqrt{1 - y^2} = 0$

    (iii) $y' = \dfrac{\cos x}{\sin y}\, , \, y(\pi) = \dfrac{\pi}{2}$

20. Find the general solution of $y' + (\cot x)\, y = x \csc x.$

21. Solve the initial value problem.

    $y' - 2xy = 1,\;\; y(0) = y_0$

22. Solve the initial value problem.

    (i) $y' + \left(\dfrac{1 + x}{x}\right) y = 0,\; y(1) = 1$

    (ii) $xy' + \left(1 + \dfrac{1}{ln\, x}\right) y = 0,\;\; y(e) = 1$

    (iii) $xy' + (1 + x \cot x)\, y = 0,\;\; y\left(\dfrac{\pi}{2}\right) = 2.$

23. Find the general solution of the following equations :

(i) $y' + \dfrac{1}{x} y = \dfrac{7}{x^2} + 3$

(ii) $y' + \dfrac{4}{x-1} y = \dfrac{1}{(x-1)^5} + \dfrac{\sin x}{(x-1)^4}$

(iii) $xy' + (1 + 2x^2) y = x^3 e^{-x^2}$

(iv) $xy' + 2y = \dfrac{2}{x^2} + 1$

(v) $y' + (\tan x) y = \cos x$

(vi) $(1 + x) y' + 2y = \dfrac{\sin x}{1 + x}$

(vii) $y' + (2 \sin x \cos x) y = e^{-\sin^2 x}$

(viii) $x^2 y' + 3xy = e^x$

24. Solve the following initial value problems :

(i) $y' + 2xy = x^2, \ y(0) = 3$

(ii) $y' + \dfrac{1}{x} y = \dfrac{\sin x}{x^2}, \ y(1) = 2$

(iii) $y' + y = \dfrac{e^{-x} \tan x}{x}, \ y(1) = 0$

(iv) $y' + \dfrac{2x}{1 + x^2} y = \dfrac{e^x}{(1 + x^2)^2}, \ y(0) = 1$

25. Prove that $y' = \dfrac{10}{3} xy^{2/5}, \ y(0) = 0$ has more than one solutions in a rectangle containing the point $(0, 0)$.

26. Show that the equation $y' = \dfrac{10}{3} xy^{2/5}, \ y(0) = -1$ has a unique solution on some open interval containing $x_0 = 0$.

27. Discuss the existence and uniqueness of the solutions of the following differential equations :

(i) $y' = \dfrac{x^2 + y^2}{\sin x}$ (ii) $y' = \dfrac{e^x + y}{x^2 + y^2}$

(iii) $y' = \tan (xy)$ (iv) $y' = \dfrac{x^2 + y^2}{\ln(x - y)}$

(v) $y' = (x^2 + y^2) y^{1/3}$ (vi) $y' = 2xy$

(vii) $y' = \ln(1 + x^2 + y^2)$ (viii) $y' = (x^2 + y^2)^{1/2}$

(ix) $y' = x(y^2 - 1)^{2/3}$

28. Consider the initial value problem
$$y' = 3x(y - 1)^{1/3}, \quad y(x_0) = y_0.$$

(a) For what points (x_0, y_0) does the equation has a solution ?

(b) For what points (x_0, y_0) does the problem has a unique solution on some open interval that contains x_0 ?

29. Solve the following differential equations :

(i) $(4x + 3y + 1) \, dx + (3x + 2y + 1) \, dy = 0$

(ii) $\dfrac{dy}{dx} = \dfrac{2x - y}{x + 2y - 5}$

(iii) $(1 + 4xy + 2y^2) \, dx + (1 + 4xy + 2x^2) \, dy = 0$

(iv) $[1 + e^{x/y}] \, dx + e^{x/y} \left[1 - \left(\dfrac{x}{y} \right) \right] dy = 0$

(v) $(y \sin 2x) \, dx - (1 + y^2 + \cos^2 x) \, dy = 0$

30. Solve the following differential equations :

(i) $(1 + y^2) \, dx = \left(\sqrt{1 + y^2} \, \sin y - xy \right) dy$

(ii) $\dfrac{dy}{dx} + \left(\dfrac{4x}{x^2 + 1} \right) y = \dfrac{1}{(x^2 + 1)^3}$

(iii) $\dfrac{dy}{dx} + \left(\dfrac{1 - 2x}{x^2} \right) y = 1$

(iv) $\dfrac{dy}{dx} + 2 \sin x = y \tan x$

(v) $x \dfrac{dy}{dx} - ay - x = 1$

(vi) Solve $\dfrac{dy}{dx} + 2y \tan x = \sin x$, Given that $y = 0$, when $x = \dfrac{\pi}{3}$.

(vii) $(1 + x^2) \dfrac{dy}{dx} + (1 - x)^2 y = x \, e^{-x} (1 + x^2)$

(viii) $(1 + y^2) \, dx + (x - e^{\tan^{-1} y}) \, dy = 0$

(ix) $\cos^2 x \dfrac{dy}{dx} + y = \tan x$

(x) Solve $\dfrac{dy}{dx} + \dfrac{y}{x} = x^2$, given that $y = 1$, when $x = 1$.

(xi) $\dfrac{dy}{dx} = \dfrac{e^x - 3xy}{x^2}$

(xii) $(1 + x^2)\dfrac{dy}{dx} + y = \tan^{-1} x$

(xiii) $(x^2 - 1)\dfrac{dy}{dx} + 2xy = \dfrac{2}{x^2 - 1}$

(xiv) $\dfrac{dy}{dx} + y \cot x = 2 \cos x$

(xv) $(x \log x)\dfrac{dy}{dx} + y = \dfrac{2}{x} \log x$

31. Solve the following differential equations :

(i) $x\dfrac{dy}{dx} + 3y = x^4 e^{1/x^2} \cdot y^3$　　　(ii) $\dfrac{dy}{dx} + \left(\dfrac{1}{x}\right) y = x^2 y^6$

(iii) $x\dfrac{dy}{dx} + \dfrac{y^2}{x} = y$　　　(iv) $x\dfrac{dy}{dx} + y = x^2 y^4$

(v) $(x^3 y^2 + xy)\, dx = dy$　　　(vi) $\dfrac{dy}{dx} + \dfrac{y}{x - 1} = xy^{1/3}$

(vii) $nx\dfrac{dy}{dx} + 2y = xy^{n+1}$

(viii) $\dfrac{dy}{dx} - 2y \tan x = y^2 \tan x$

Answers

[A] (1) False　(2) False　(3) False　(4) True　(5) True

[B] (1) - (b)　(2) - (b)　(3) - (a)　(4) - (c)　(5) - (a)　(6) - (a)　(7) - (c)
　　(8) - (a)

[D]

1. (i) $n = 4$,　(ii) $n = -12$,　(iii) $n = 0$,　(iv) $n = \dfrac{1}{12}$,　(v) $n = 3$,

　　(vi) $n = 4$

2. (i) $\dfrac{\partial u}{\partial x} = \dfrac{x^2 + 2xy - y^2}{(x + y)^2 + (x^2 + y^2)^2}$

　　$\dfrac{\partial u}{\partial y} = \dfrac{y^2 + xy - x^2}{(x + y)^2 + (x^2 + y^2)^2}$

　　(ii) $\dfrac{\partial u}{\partial x} = -\dfrac{(1 + y^2)}{(x + y)^2}$ and $\dfrac{\partial u}{\partial y} = -\dfrac{(1 + x^2)}{(x + y)^2}$

　　(iii) $\dfrac{\partial u}{\partial x} = \dfrac{2x}{x^2 + y^2}$ and $\dfrac{\partial u}{\partial y} = \dfrac{2y}{x^2 + y^2}$

3. (i) $(1 + x^2 + y^2)^{-3/2}$

(ii) 0

5. (i) Order = 1 and degree = 2

(ii) Order = 2 and degree = 3

(iii) **Hint :** We can write the given equation as

$$\frac{dy}{dx} = \frac{x^3 - y^3}{x^2 y - xy^2}$$

∴ Order = 1 and degree = 1

(iv) **Hint :** $y = px \pm c\sqrt{p^2 + 1} \Rightarrow y - px = \pm c\sqrt{p^2 + 1}$
    Ans. : Order = 1 and degree = 2

(v) Order = 2 and degree = 2

(vi) Order = 2 and degree is not defined.

6. **Hint :** The general equation of the family of all straight lines making equal intercepts c on the axes is

$$\frac{x}{c} + \frac{y}{c} = 1 \Rightarrow x + y = c$$

Ans. : $1 + \dfrac{dy}{dx} = 0$

7. **Hint :** The general equation of the family of all the circles having their centes (a, 0) on x-axis and radii equal to unity is

$$(x - a)^2 + (y - 0)^2 = 1 \Rightarrow (x - a)^2 + y^2 = 1$$

Ans. : $y^2 \left(\dfrac{dy}{dx}\right)^2 + y^2 = 1$

8. **Ans. :** $\dfrac{d^2 y}{dx^2} - y = 0$

9. **Ans. :** $\dfrac{d^2 y}{dx^2} - 2\dfrac{dy}{dx} + 2y = 0$

10. (i) $y = x \log (x + 1) - x + \log (x + 1) + c$

(ii) $\log (y) = \dfrac{2}{3} x^3 + x^2 + 2x + 2 \log (x - 1) + c$

(iii) $\dfrac{1}{x} + \dfrac{1}{y} + \log (y) - \log (x) = c,$ where $k = -c$

(iv) $y + \dfrac{2}{3} (a + x)^{3/2} - a(a + x)^{1/2} = c$

(v) $(1 + \sin x) (1 + \cos y) = c,$

(vi) $y^2 \log y = e^x \sin^2 x + c$

(vii) $(1 + \log x)^2 + 2 \tan^{-1} y = 1 + \dfrac{\pi}{2}$

(viii) $y = e^{\sin^2 x}$

(ix) $\cot y = 2\,(1 + x^2)$

(x) $y = \dfrac{a}{2} \log \left| \dfrac{x - y - a}{x - y + a} \right| + c$

(xi) Put $x + y = v$ and proceed

Ans. : $y - \tan\left(\dfrac{x + y}{2}\right) = c$

(xii) Put $x + y = v$ and proceed

Ans. : $1 + \tan\left(\dfrac{x + y}{2}\right) = ce^x$

11. (i) $e^{\frac{x^2}{2y^2}} = cy$

(ii) $2 \log (y) + 3 \log ((2x + y)(2x - y)) = c$

(iii) $-\dfrac{x^3}{3y^3} + \log (y) = c$

(iv) $x^2(2xy + y^2) = c$

(v) $x\sqrt{\dfrac{y}{y + 2x}} = \pm c$

(vi) $\log (y) + \dfrac{x}{x + y} = c$

(vii) $\sqrt{x^2 + y^2} + x \log x = cx$

(viii) $y^2 - x^2 = c^2\,(y^2 + x^2)^2$

12. (i) $(x + y - 2) = c^2\,(x - y)^3$

(ii) $(y - x + 3)^4 = c(2x + y - 3)$

(iii) $x - 2y + \log (x - y + 2) = c$

(iv) $2 \tan^{-1}\left\{\dfrac{(2y + 1)}{(2x + 1)}\right\} = \log\left\{c^2\left(x^2 + y^2 + x + y + \dfrac{1}{2}\right)\right\}$

(v) $(y - 2x)^2 = c(x + 2y - 5)$

14. $y = \dfrac{-1}{x^2 + c}$

15. $y = \dfrac{1 + ce^{-x^2/2}}{1 - ce^{-x^2/2}}$

20. $y = \dfrac{x^2}{2\sin x} + \dfrac{c}{\sin x}$

21. $y = e^{x^2}\left(c + \int e^{-x^2}\,dx\right)$

29. (i) $2x^2 + 3xy + y^2 + x + y = c$

    (ii) $x^2 - y^2 - xy + 5y = c$

    (iii) $x + 2x^2y + 2xy^2 + y = c$

    (iv) $x + ye^{x/y} = c$

    (v) $y\cos 2x + 2y + \dfrac{2}{3}y^3 = c$

30. (i)   **Hint :** $\text{I.F.} = \sqrt{1 + y^2}$

    Ans. : $\cos y + x\sqrt{1 + y^2} = c$

    (ii)   **Hint :** $\text{I.F.} = (x^2 + 1)^2$

    Ans. : $y(x^2 + 1)^2 = \tan^{-1}x + c$

    (iii)   **Hint :** $\text{I.F.} = \dfrac{e^{-\frac{1}{x}}}{x^2}$

    Ans. : $ye^{-\frac{1}{x}} = x^2 e^{-\frac{1}{x}} + cx^2$

    (iv)   **Ans. :** $y\cos x = \dfrac{1}{2}\cos 2x + c$

    (v) $y = \dfrac{x}{1-a} - \dfrac{1}{a} + cx^a$

    (vi) $y = \cos x - 2\cos^2 x$

    (vii) $\dfrac{ye^x}{1 + x^2} = \dfrac{1}{2}\log(1 + x^2) + c$

    (viii) **Hint :** $\text{I.F.} = e^{\tan^{-1}y}$

    Ans. : $xe^{\tan^{-1}y} = \dfrac{1}{2}e^{2\tan^{-1}y} + c$

    (ix)   **Ans. :** $y = \tan x - 1 + ce^{-\tan x}$

    (x)   **Ans. :** $4xy = x^4 + 3$

    (xi)   **Ans. :** $x^3 y = (x - 1)e^x + c$

    (xii) **Hint :** $\text{I.F.} = e^{\tan^{-1}x}$

    Ans. : $y = (\tan^{-1}x - 1) + ce^{-\tan^{-1}x}$

    (xiii) **Hint :** $\text{I.F.} = x^2 - 1$

       Ans. : $y(x^2 - 1) = \log\left(\dfrac{x - 1}{x + 1}\right) + c$

(xiv) **Hint :** I.F. $= \sin x$

        Ans. : $y \sin x = \dfrac{1}{2} \cos 2x + c$

(xv) **Hint :** I.F. $= \log x$

        Ans. : $y \log x + \dfrac{2}{x}(\log x + 1) + c$

31. (i) **Hint :** Bernoulli's equation

        Ans. : $y^2 x^6 \left(e^{\frac{1}{x^2}} + c\right) = 1$

  (ii) **Hint :** Bernoulli's equation

        Ans. : $\dfrac{1}{y^5} = \dfrac{5}{2} x^3 + c\, x^5$

 (iii) **Hint :** Bernoulli's equation

        Ans. : $x = y \log (cx)$

 (iv) **Hint :** Bernoulli's equation

        Ans. : $\dfrac{1}{y^3} = x^2 (3 - cx)$

  (v) **Hint :** Bernoulli's equation

        Ans. : $2y \left(1 - \dfrac{x^2}{2}\right) - cy\, e^{-\frac{x^2}{2}}$

 (vi) **Hint :** Bernoulli's equation

        Ans. : $y^{\frac{2}{3}} (x - 1)^{\frac{2}{3}} = \dfrac{2}{5} x(x - 1)^{\frac{5}{3}} - \dfrac{3}{20} (x - 1)^{\frac{8}{3}} + c$

(vii) **Hint :** Bernoulli's equation

        Ans. : $\dfrac{1}{xy^n} = cx + 1$

(viii) **Hint :** Bernoulli's equation

        Ans. : $\sec^2 x + \dfrac{1}{3} y \tan^3 x + cy = 0$

☞ ☞ ☞

Chapter 4...
Exact Differential Equations

Jean-Baptiste Le Rond d'Alembert

Jean-Baptiste Le Rond d'Alembert (16 November 1717 - 29 October 1783) was a French mathematician, **mechanician, physicist**, philosopher, and **music theorist**. Until 1759 he was co-editor with **Denis Diderot** of the **Encyclopedie. D'Alembert's formula** for obtaining solutions to the **wave equation** is named after him. The wave equation is sometimes referred to as **d'Alembert's equation.**

In 1743, he published his most famous work, Traite de dynamique, in which he developed his own **laws of motion.**

In 1752, he wrote about what is now called **D'Alembert's paradox**: that the **drag** on a body immersed in an **inviscid, incompressible fluid** is zero.

4.1 Exact Differential Equations

4.1.1 Introduction

Given a differential equation $y' = f(x, y)$, we write it as

$$M(x, y)\, dx + N(x, y)\, dy = 0.$$

This equation can be interpreted as $M(x, y) + N(x, y) \dfrac{dy}{dx} = 0$, where x is the independent variable and y is the dependent variable. Implicit solution of this equation is given by $u(x, y) = c$. Every differentiable function $y = y(x)$, that satisfies $u(x, y) = c$ is a solution of this differential equation.

Example : If $u(x, y) = x^4 y^3 + x^2 y^5 + 2xy = c$ is an implicit solution of $(4x^3 y^3 + 2xy^5 + 2y)\, dx + (3x^4 y^2 + 5x^2 y^4 + 2x)\, dy = 0$

Definition : The differential equation, obtained from its primitive (solution) by differentiating the solution without any further operation of elimination or reduction is called an **exact differential equation** or alternatively; the differential equation M dx + N dy = 0 is said to be an exact differential equation, if there exists a function u(x, y) such that

$$M\ dx + N\ dy\ =\ du$$

For example, consider the solution

$$x^2 y^3\ =\ c \qquad\qquad \dots \text{(i)}$$

Differentiating equation (i) w.r.t. x, we get

$$2xy^3 + 6x^2 y^2\ \frac{dy}{dx}\ =\ 0$$

$$\Rightarrow \qquad 2xy^3\ dx + 6x^2 y^2\ dy\ =\ 0 \qquad\qquad \dots \text{(ii)}$$

Cancelling the common factor xy^2, throughout the equation (ii), we get

$$2y\ dx + 6x\ dy\ =\ 0 \qquad\qquad \dots \text{(iii)}$$

Here equation (ii) is obtained by directly differentiating the relation (i) without cancelling the common factor involved in the equation, such a equation (ii) is called an exact **differential equation**.

But the equation (iii) is obtained from equation (ii) by cancelling the common factor xy^2 throughout the equation (ii). This equation (iii) will not be called an **exact equation**. We note here that both the equations (ii) and (iii) have got the same solution $x^2 y^3 = c$, but these two equations (ii) and (iii) have different forms.

Naturally, we see for the conditions under which the differential equation will be having exact form i.e. we find in what way M and N are placed in the differential equation M dx + N dy = 0, so that M dx + N dy = 0 assumes an exact form.

| **Theorem 1** | If the function u = u(x, y) has continuous partial derivatives u_x and u_y, then u(x, y) = c is an implicit solution of the differential equation

$$u_x\ (x, y)\ dx + u_y\ (x, y)\ dy = 0.$$

Proof : Suppose, y is a function of x.

Differentiating u(x, y) = c with respect to x, we get,

$$\frac{\partial u}{\partial x}\ (x, y) + \frac{\partial u}{\partial y}\ (x, y), \frac{dy}{dx}\ =\ 0$$

$$\Rightarrow \qquad u_x(x, y) + u_y(x, y)\ \frac{dy}{dx}\ =\ 0$$

$$\Rightarrow \qquad u_x(x, y)\ dx + u_y(x, y)\ dy\ =\ 0$$

Hence, u(x, y) = c satisfies the differential equation

$$M(x, y)\, dx + N(x, y)\, dy = 0,$$

where, $M(x, y) = \dfrac{\partial u}{\partial x}$, $N(x, y) = \dfrac{\partial u}{\partial y}$. ∎

4.1.2 Necessary and Sufficient Condition for Exactness

$\boxed{\textbf{Theorem 2}}$ **(Necessary and Sufficient Condition) :**

The necessary and sufficient condition that the differential equation M dx + N dy = 0 to be exact is

$$\frac{\partial M}{\partial y} = \frac{\partial N}{\partial x}, \; \forall (x, y) \text{ in the domain.}$$

Proof : (1) The condition is necessary :

Let the equation M dx + N dy = 0   ... (i)
be exact.

∴ By definition, this equation must have been derived from its solution u(x, y) = c, say by direct differentiation without any further operation of elimination or reduction.

So differentiating u(x, y) = c, we get

$$du = 0$$

$$\Rightarrow \quad \frac{\partial u}{\partial x} dx + \frac{\partial u}{\partial y} dy = 0 \,...\,(ii) \quad \left(\because \; du = \frac{\partial u}{\partial x} dx + \frac{\partial u}{\partial y} dy \right)$$

which should be same as M dx + N dy = 0, if the equation is exact.

On comparing the coefficients of dx and dy in equation (i) and (ii), we get

$$\frac{\partial u}{\partial x} = M \;\text{ and }\; \frac{\partial u}{\partial y} = N$$

Then

$$\frac{\partial}{\partial y}\left(\frac{\partial u}{\partial x}\right) = \frac{\partial M}{\partial Y} \;\text{ and }\; \frac{\partial}{\partial x}\left(\frac{\partial u}{\partial y}\right) = \frac{\partial N}{\partial x}$$

$$\Rightarrow \qquad \frac{\partial^2 u}{\partial y\, \partial x} = \frac{\partial M}{\partial y} \;\text{ and }\; \frac{\partial^2 y}{\partial x\, \partial y} = \frac{\partial N}{\partial x}$$

But

$$\frac{\partial^2 u}{\partial y\, \partial x} = \frac{\partial^2 u}{\partial x\, \partial y} \text{ and hence, we have}$$

$$\frac{\partial M}{\partial y} = \frac{\partial N}{\partial x}$$

Thus if the equation M dx + N dy = 0 is exact, then $\dfrac{\partial M}{\partial y} = \dfrac{\partial N}{\partial x}$.

(2) The condition is sufficient :

Let $\dfrac{\partial M}{\partial y} = \dfrac{\partial N}{\partial x}$ for the differential equation $M\,dx + N\,dy = 0$.

We have to show that the equation

$$M\,dx + N\,dy = 0 \text{ is exact differential equation}$$

For this, let $\qquad \int M\,dx = V$

then $\qquad \dfrac{\partial V}{\partial x} = M \qquad\qquad\qquad \dots \text{(iii)}$

Here we have partial differentiation w.r.t. x treating y constant.

$\Rightarrow \qquad \dfrac{\partial}{\partial y}\left(\dfrac{\partial u}{\partial x}\right) = \dfrac{\partial M}{\partial y}$

$\Rightarrow \qquad \dfrac{\partial^2 V}{\partial y\,\partial x} = \dfrac{\partial M}{\partial y}$

$\qquad\qquad\qquad = \dfrac{\partial N}{\partial x} \qquad\qquad \because \dfrac{\partial M}{\partial y} = \dfrac{\partial N}{\partial x} \text{ given}$

$\Rightarrow \qquad \dfrac{\partial N}{\partial x} = \dfrac{\partial^2 V}{\partial x\,\partial y} \qquad\qquad \because \dfrac{\partial^2 V}{\partial y\,\partial x} = \dfrac{\partial^2 V}{\partial x\,\partial y}$

$\Rightarrow \qquad \dfrac{\partial}{\partial x}(N) = \dfrac{\partial}{\partial x}\left(\dfrac{\partial V}{\partial y}\right)$

$\Rightarrow \qquad \dfrac{\partial}{\partial x}(N) - \dfrac{\partial}{\partial x}\left(\dfrac{\partial V}{\partial y}\right) = 0$

$\Rightarrow \qquad \dfrac{\partial}{\partial x}\left(N - \dfrac{\partial V}{\partial y}\right) = 0$

Integrating w.r.t. x treating y-constant, we get

$$N - \dfrac{\partial V}{\partial y} = \text{Constant of integration}$$

$\Rightarrow \qquad N = \dfrac{\partial V}{\partial y} + \text{Constant of integration} \dots \text{(iv)}$

But since we are treating y as a constant quantity. Hence, y may appear as a constant in the above constant of integration and hence we write this constant of integration as a function of y say $\phi'(y)$, where $\phi'(y)$ is the derivative of $\phi(y)$ w.r.t. y.

$\therefore$ From equation (iv), we have

$$N = \frac{\partial v}{\partial y} + \phi'(y)$$

Thus, from equation (iii) and (iv), we have

$$M = \frac{\partial V}{\partial x} \text{ and } N = \frac{\partial V}{\partial y} + \phi'(y)$$

Then

$$M \, dx + N \, dy \equiv \frac{\partial V}{\partial x} \, dx + \left[\frac{\partial V}{\partial y} + \phi'(y)\right] dy$$

$$\equiv \frac{\partial V}{\partial x} \, dx + \frac{\partial V}{\partial y} \, dy + \phi'(y) \, dy$$

$$\equiv dV + d(\phi(y))$$

$$\equiv d[V + \phi(y)]$$

$$M \, dx + N \, dy \equiv d[V + \phi(y)]$$

Hence, $M \, dx + N \, dy = 0$ is exact differential equation.

Hence proof.

Working Rule for Solving an Exact Differential Equation :

We have seen from above that, if $\dfrac{\partial M}{\partial y} = \dfrac{\partial N}{\partial x}$, then $M \, dx + N \, dy = 0$ is exact differential equation and further

$$M \, dx + N \, dy = d \, [V + \phi(y)]$$

where, V and $\phi(y)$ are already defined. Hence integrating the equation, we obtain

$$V + \phi(y) = c \qquad \qquad \text{... (i)}$$

where, c is constant as a solution of $M \, dx + N \, dy = 0$

Further $V + \phi(y) = c$ involves only one constant.

Hence, $V + \phi(y) = c$ is the general solution of $M \, dx + N \, dy = 0$

Here v by definition is $\displaystyle\int M \, dx$ i.e.

y-constant

$$V = \int_{\text{y-constant}} M \, dx \qquad \qquad \text{... (ii)}$$

and $\phi(y) = \int \phi'(y) \, dy$, where $\phi'(y)$ by definition involves the terms in N which are functions of y alone. [i.e. terms in N free from x]

Thus we get a rule to get the general solution (i) of exact differential equation M dx + N dy as

$$\int\limits_{\text{y-constant}} M\ dx \ + \int (\text{Terms in N free from x})\ dy = c.$$

If M dx + N dy = 0 is an exact equation $\left(\dfrac{\partial M}{\partial y} = \dfrac{\partial N}{\partial x} \text{ is satisfied} \right)$ then

its general solution is:

   (i)   Integrate M w.r.t. x treating y constant.

   (ii) Integrate those terms in N that are free from x w.r.t. y.

   (iii) Equate the sum of equation (i) and (ii) to an arbitrary constant.

In short, if the differential equation M dx + N dy = 0 is exact, then its general solution is

$$\int\limits_{\text{y-constant}} M\ dx \ + \int (\text{Terms in N free from x})\ dy = c.$$

Illustrative Examples

Example 4.1 : *Solve $(4x^3y^3 + 3x^2)\ dx + (3x^4y^2 + 6y^2)\ dy.$*

Solution : Here, $M(x, y) = 4x^3y^3 + 3x^2$ and $N(x, y) = 3x^4y^2 + 6y^2$

$$\Rightarrow \quad \frac{\partial M}{\partial y} = 12x^3y^3 \ \text{ and } \ \frac{\partial N}{\partial x} = 12x^3y^2$$

$$\Rightarrow \quad \frac{\partial M}{\partial y} = \frac{\partial N}{\partial x}, \text{ hence, the equation is exact.}$$

Now, if $u(x, y)$ is a function such that $du = Mdx + Ndy$, then

$$\frac{\partial u}{\partial x}\ dx + \frac{\partial u}{\partial y}\ dy = Mdx + Ndy \ \Rightarrow \ \frac{\partial u}{\partial x} = M, \ \frac{\partial u}{\partial y} = N.$$

Integrating $\dfrac{\partial u}{\partial x} = M$ with respect to x, we get

$$u(x, y) \ = \ \int Mdx + g(y) \qquad\qquad \dots (1)$$

where, g is some function independent of x which may depend on y.

To determine the unknown function g(y), differentiate (1) with respect to y.

From equation (1), we have

$$u(x, y) \ = \ \int (4x^3y^3 + 3x^2)\ dx + g(y)$$

$$u(x, y) \ = \ x^4y^3 + x^3 + g(y)$$

$$\therefore \qquad \frac{\partial u}{\partial y} = 3x^4 y^2 + g'(y) \qquad\qquad \dots (2)$$

But $\dfrac{\partial u}{\partial y} = N = 3x^4 y^2 + 6y^2$, so that equation (2) becomes

$$3x^4 y^2 + 6y^2 = 3x^4 y^2 + g'(y)$$

$$\Rightarrow \quad g'(y) = 6y^2 \quad \Rightarrow \quad g(y) = \int 6y^2 \, dy + c \quad \Rightarrow \quad g(y) = 2y^3 + c$$

Hence, $\qquad\qquad u(x, y) = x^4 y^3 + x^3 + 2y^3 + c$

Thus the general solution is

$$x^4 y^3 + x^3 + 2y^3 + c = 0$$

Example 4.2 : *Solve,* $(1 + 6y^2 - 3x^2 y)\dfrac{dy}{dx} = 3xy^2 - x^2.$

Solution : The given equation can be written as

$$(3xy^2 - x^2)\, dx + (3x^2 y - 6y^2 - 1)\, dy = 0$$

Comparing this equation with M dx + N dy = 0, we get

$$M = 3xy^2 - x^2 \ \text{ and } \ N = 3x^2 y - 6y^2 - 1$$

$$\Rightarrow \quad \frac{\partial M}{\partial y} = 6xy \qquad \text{and} \quad \frac{\partial N}{\partial x} = 6xy$$

$$\Rightarrow \quad \frac{\partial M}{\partial y} = \frac{\partial N}{\partial x}$$

$\therefore$ The given differential equation is exact.

Hence, the solution is

$$\int_{\text{y-constant}} M \, dx \ + \int (\text{Terms in N free from x}) \, dy = c$$

$$\Rightarrow \quad \int_{\text{y-constant}} (3xy^2 - x^2)\, dx + \int (-6y^2 - 1) \, dy = c$$

$$\Rightarrow \quad \frac{3}{2} x^2 y^2 - \frac{x^3}{3} - 2y^3 - y = c$$

which is the required general solution.

Example 4.3 : *Solve,* $(x^2 - 2xy - y^2)\, dx - (x + y)^2 \, dy = 0.$

Solution : The given differential equation is

$$(x^2 - 2xy - y^2)\, dx - (x^2 + 2xy + y^2)\, dy = 0$$

Comparing this equation with $M\,dx + N\,dy = 0$, we get

$$M = x^2 - 2xy - y^2 \text{ and } N = -(x^2 + 2xy + y^2)$$

$$\Rightarrow \quad \frac{\partial M}{\partial y} = -2x - 2y \quad \text{and} \quad \frac{\partial N}{\partial x} = -2x - 2y$$

$$\Rightarrow \quad \frac{\partial M}{\partial y} = \frac{\partial N}{\partial x}$$

$\Rightarrow$ The given differential equation is exact.

Hence, its general solution is

$$\int_{y\text{-constant}} M\,dx \; + \int (\text{Terms in N free from x})\,dy = c$$

$$\Rightarrow \quad \int_{y\text{-constant}} (x^2 - 2xy - y^2)\,dx + \int - y^2\,dy = c$$

$$\Rightarrow \quad \frac{1}{3}x^3 - x^2 y - xy^2 - \frac{1}{3}y^3 = c$$

which is the required general solution.

Note : The given differential equation is homogeneous differential equation. Hence it can also be solved by putting $y = vx$.

Proposition : Suppose that $M(x, y)$, $N(x, y)$ are functions such that M, N, $\dfrac{\partial y}{\partial x}, \dfrac{\partial M}{\partial y}, \dfrac{\partial N}{\partial x}, \dfrac{\partial N}{\partial y}$ are continuous on an open rectangle R. Let $G(x)$ be a function such that $G'(x) = M$. If $M_y \neq N_x$ in R, then the function

$$N - \frac{\partial G}{\partial y} \text{ is not independent of x.}$$

Proof : Since, G is an antiderivative of M,

$$\frac{\partial G}{\partial x} = M \Rightarrow G = \int \frac{\partial G}{\partial x} = \int M dx$$

Hence, $$N - \frac{\partial G}{\partial y} = N - \frac{\partial}{\partial y}(\int M dx)$$

Now, $$\frac{\partial}{\partial x}\left(N - \frac{\partial G}{\partial y}\right) = \frac{\partial}{\partial x}\left(N - \frac{\partial}{\partial y}(\int M dx)\right)$$

$$= \frac{\partial N}{\partial x} - \frac{\partial}{\partial x}\left(\frac{\partial}{\partial y}(\int M dx)\right)$$

$$= \frac{\partial N}{\partial x} - \frac{\partial}{\partial y}\left(\frac{\partial}{\partial x}\int M dx\right)$$

$$= \frac{\partial N}{\partial x} - \frac{\partial M}{\partial y} \neq 0, \text{ since } M_y \neq N_x.$$

Thus, $\dfrac{\partial}{\partial x}\left(N - \dfrac{\partial G}{\partial y}\right) \neq 0 \Rightarrow N - \dfrac{\partial G}{\partial y}$ is not independent of x.

Proposition : If the equation $M_1 dx + N_1 dy = 0$ and $M_2 dx + N_2 dy = 0$ are exact, then the equation $(M_1 + M_2)\, dx + (N_1 + N_2)\, dy = 0$ is exact.

Proof : Since, $M_1 dx + N_1 dy = 0$ is exact, $\exists\ u_1(x, y)$ such that

$$du_1 = M_1 dx + N_1 dy.$$

Also, $M_2 dx + N_2 dy = 0$ is exact, $\exists\ u_2(x, y)$ such that

$$du_2 = M_2 dx + N_2 dy.$$

Now, let $u = u_1 + u_2$.

Then $du = du_1 + du_2 = (M_1 + M_2)\, dx + (N_1 + N_2)\, dy.$

$\therefore \quad (M_1 + M_2)\, dx + (N_1 + N_2)\, dy = 0$ is exact.

Proposition : Suppose that $A(x, y)$, $B(x, y)$, $C(x, y)$ and $D(x, y)$ are functions such that $x\dfrac{\partial A}{\partial y} + y\dfrac{\partial B}{\partial y} + B = x\dfrac{\partial C}{\partial x} + y\dfrac{\partial D}{\partial x} + C.$

Then the differential equation,

$$(Ax + By)\, dx + (Cx + Dy)\, dy = 0 \text{ is exact.}$$

Proof : Left as an exercise.

Illustrative Example

Example 4.4 : *Find the condition on the constants P, Q, R, S, T and U such that the equation*

$$(Px^2 + Qxy + Ry^2)\, dx + (Sx^2 + Txy + Uy^2)\, dy = 0 \text{ is exact.}$$

Solution : Left as an exercise.

Proposition : Suppose all second order partial derivatives of

$M = M(x, y)$ and $N = N(x, y)$ are continuous and $Mdx + Ndy = 0,$ $-Ndx + Mdy = 0$ are exact on an open rectangle R. Then

$$M_{xx} + M_{yy} = N_{xx} + N_{yy} = 0 \text{ on R.}$$

Proof : Since, $Mdx + Ndy = 0$ and $-Ndx + Mdy = 0$ are exact, we have $M_y = N_x$ and $-N_y = M_x$ on R.

$\Rightarrow \quad M_{yy} = N_{xy}$ and $-N_{yx} = M_{xx} \Rightarrow M_{xx} + N_{yy} = 0.$

Also, $M_{yx} = N_{xx}$ and $-N_{yy} = M_{xy} \Rightarrow N_{xx} + N_{yy} = 0.$

Remark : A function $f(x, y)$ with continuous second order partial derivatives is said to be harmonic if $f_{xx} + f_{yy} = 0 \; \forall \; (x, y)$.

Proposition : Suppose that all second order partial derivatives of $u(x, y)$ are continuous and u is harmonic function on an open rectangle R. Then the differential equation

$$- u_y dx + u_x dy = 0 \text{ is exact on R.}$$

Moreover, there exists a function $v(x, y)$ such that $v_x = - u_y$ and $v_y = u_x$ on R.

Proof : Since, u is harmonic on R, we have

$$u_{xx} + u_{yy} = 0 \;\; \text{i.e.} \;\; \frac{\partial^2 u}{\partial x^2} + \frac{\partial^2 u}{\partial y^2} = 0$$

$$\Rightarrow \quad u_{xx} = - u_{yy} \text{ on R.}$$

$$\Rightarrow \quad \frac{\partial u_x}{\partial x} = \frac{\partial}{\partial y}(-u_y) \text{ on R.}$$

Thus, if $M = -u_y$ and $N = u_x$, then the equation $Mdx + Ndy = 0$ is exact.

Hence, the equation $-u_y dx + u_x dy = 0$ is exact.

Further, there exists a function $v(x, y)$ such that

$$dv = - u_y dx + u_x dy, \text{ so that}$$

$$v_x = \frac{\partial v}{\partial x} = -u_y \text{ and } v_y = \frac{\partial v}{\partial y} = u_x$$

Note : Here the function $v(x, y)$ is called as the harmonic conjugate of $u(x, y)$ on R.

Illustrative Example

Example 4.5 : *Prove that the function $u(x, y) = x^2 - y^2$ is harmonic. Find harmonic conjugate of u.*

Solution : Here $u_x = \dfrac{\partial^2 u}{\partial x^2} = 2$ and $u_{yy} = \dfrac{\partial^2 u}{\partial y^2} = -2$.

$$\Rightarrow \quad u_{xx} + u_{yy} = 2 - 2 = 0, \; \forall \; (x, y).$$

$$\therefore \quad u \text{ is harmonic on } \mathbb{R}^2.$$

Next to find harmonic conjugate v and u, we have $u_x = v_y$ and $u_y = -v_x$.

$$\Rightarrow \quad v_y = 2x \text{ and } v_x = 2y$$

$\Rightarrow$ Integrating first equation w.r.t y, we get,

$$v = 2xy + \phi(x), \text{ where } \phi \text{ is a function of x independent of y.}$$

Now differentiating w.r.t. x, we get,

$$v_x = 2y + \phi(x)$$

$$\Rightarrow \quad 2y = 2y + \phi'(x)$$

$$\Rightarrow \quad 2y = 2y + \phi'(x) \quad \Rightarrow \quad \phi'(x) = 0 \quad \Rightarrow \quad \phi(x) = c.$$

Hence, $v(x, y) = 2xy + c$

4.2 Integrating Factor

Definition : A differential equation which is not an exact differential equation, can be reduced to an exact equation by multiplying the equation by some function of x and y. Say $\mu(x, y)$, then such a function $\mu(x, y)$ is called an **integrating factor** (I.F.) of the differential equation.

For example, consider an equation

$$y\, dx - x\, dy\ =\ 0 \qquad \text{... (i)}$$

Comparing the equation (i), with M dx + N dy = 0, we get

$$M = y \quad \text{and} \quad N = -x$$

$$\Rightarrow \quad \frac{\partial M}{\partial y} = 1 \quad \text{and} \quad \frac{\partial N}{\partial x} = -1$$

$$\therefore \quad \frac{\partial M}{\partial y} \neq \frac{\partial N}{\partial x}$$

Hence the given equation is not exact equation. It will be shown that this equation (i) can be reduced into an exact equation by multiplying the equation (i) by $\dfrac{1}{xy}$ or $\dfrac{1}{y^2}$ or $\dfrac{1}{x^2}$ which are called integrating factors.

Multiplying the equation (i) by I.F. $\dfrac{1}{xy}$, we get

$$\left(\frac{1}{x}\right) dx - \left(\frac{1}{y}\right) dy\ =\ 0 \qquad \text{... (ii)}$$

Here $M = \dfrac{1}{x}$ and $N = -\dfrac{1}{y}$

$$\Rightarrow \quad \frac{\partial M}{\partial y} = 0 \quad \text{and} \quad \frac{\partial N}{\partial x} = 0$$

$$\Rightarrow \quad \frac{\partial M}{\partial y} = \frac{\partial N}{\partial x}$$

$\Rightarrow \quad$ The equation (ii) is exact equation

Hence its general solution is

$$\int\limits_{\substack{y\text{-constant}}} \frac{1}{x}\, dx + \int \left(-\frac{1}{y}\right) dy = c_1$$

$\Rightarrow \qquad\qquad \log(x) - \log(y) = c_1$

$\Rightarrow \qquad\qquad \log\left(\frac{x}{y}\right) = \log c$

$\Rightarrow \qquad\qquad \dfrac{x}{y} = c$

$\Rightarrow \qquad\qquad x = cy$

which is the general solution of given equation (i).

Similarly by using integrating factors $\dfrac{1}{y^2}$ and $\dfrac{1}{x^2}$, we get same solution.

Remark :

(1) The number of integrating factors for the equation

$M\, dx + N\, dy = 0$ are infinite.

(2) If μ is an integrating factor of the equation $M\, dx + N\, dy = 0$, then $\mu(M\, dx + N\, dy) = 0$ is an exact differential equation.

(3) Although an equation of the form

$$M\, dx + N\, dy = 0 \quad \text{always has integrating factor.}$$

There is no general method of finding integrating factor.

Here, we explain some rules for finding I.F. :

By Inspection : The following list of exact differentials should be noted very carefully.

(i) $\quad d\left(\dfrac{x}{y}\right) = \dfrac{y\, dx - x\, dy}{y^2}$

(ii) $\quad d\left(\dfrac{y}{x}\right) = \dfrac{x\, dy - y\, dx}{x^2}$

(iii) $\quad d\left(\tan^{-1}\dfrac{y}{x}\right) = \dfrac{x\, dy - y\, dx}{x^2 + y^2}$

(iv) $\quad d\left(\tan^{-1}\dfrac{x}{y}\right) = \dfrac{y\, dx - x\, dy}{x^2 + y^2}$

(v) $\quad d\left(\log \dfrac{x}{y}\right) = \dfrac{y\,dx - x\,dy}{xy}$

(vi) $\quad d\left(\log \dfrac{y}{x}\right) = \dfrac{x\,dy - y\,dx}{xy}$

(vii) $\quad d\,(xy) = x\,dy + y\,dx$

(viii) $d\left(\dfrac{1}{xy}\right) = -\left[\dfrac{x\,dy + y\,dx}{x^2 y^2}\right]$

(ix) $\quad d\dfrac{y^2}{x^2} = \dfrac{2xy\,dx - x^2 dy}{y^2}$

(x) $\quad d\left(\dfrac{y^2}{x}\right) = \dfrac{2xy\,dy - y^2 dx}{x^2}$

(xi) $\quad d\left(\dfrac{y^2}{x^2}\right) = \dfrac{2x^2 y\,dy - 2y^2 x\,dx}{x^4}$

(xii) $\quad d\left(\dfrac{x^2}{y^2}\right) = \dfrac{2xy^2 dx - 2yx^2 dy}{y^4}$

(xiii) $d\left(\dfrac{e^x}{y}\right) = \dfrac{ye^x\,dx - e^x\,dy}{y^2}$

(xiv) $d[\log (x^2 + y^2)] = \dfrac{2x\,dx + 2y\,dy}{x^2 + y^2}$

Illustrative Example

Example 4.6 : *Solve, $(1 + xy)y\,dx + (1 - xy)x\,dy = 0$.*

Solution : The given equation is

$$(1 + xy)y\,dx + (1 - xy)x\,xy = 0$$

Comparing the equation with $M\,dx + N\,dy = 0$, we get

$$M = (1 + xy)y \quad \text{and} \quad N = (1 - xy)x$$

$$\Rightarrow \quad \dfrac{\partial M}{\partial y} = 1 + 2xy \quad \text{and} \quad \dfrac{\partial N}{\partial x} = 1 - 2xy$$

$$\therefore \quad \dfrac{\partial M}{\partial y} \neq \dfrac{\partial N}{\partial x}$$

$\therefore$ Given equation is not exact. But we can make it exact by multiplying by I.F. We find I.F. by inspection method.

The terms of the given equation can be rearranged as

$$y\,dx + x\,dy + xy^2 dx - x^2 y\,dy = 0 \qquad \ldots (i)$$

Now $y\,dx + x\,dy$ is exact. Since $d(xy) = y\,dx + x\,dy$.

For the remaining part multiply by $\dfrac{1}{x^2 y^2}$, then the equation (i) becomes,

$$\frac{y\,dx + x\,dy}{x^2 y^2} + \frac{xy^2 dx - x^2 y\,dy}{x^2 y^2} = 0$$

$$\Rightarrow \qquad \frac{y\,dx + x\,dy}{x^2 y^2} - \frac{x\,dy - y\,dx}{xy} = 0$$

$$\Rightarrow \qquad -d\left(\frac{1}{xy}\right) - d\left[\log\left(\frac{y}{x}\right)\right]$$

$$\Rightarrow \qquad -\frac{1}{xy} - \log\left(\frac{y}{x}\right) = c_1$$

$$\Rightarrow \qquad \frac{1}{xy} + \log\left(\frac{y}{x}\right) = c_1$$

$$\log e^{1/xy} + \log\left(\frac{y}{x}\right) = \log c$$

$$\Rightarrow \qquad \log\left[\frac{y}{x}\, e^{\frac{1}{xy}}\right] = \log c$$

$$\Rightarrow \qquad \frac{y}{x}\, e^{\frac{1}{xy}} = c$$

$$\Rightarrow \qquad y\, e^{\frac{1}{xy}} = cx$$

which is required solution.

There are some rules which are useful for determining integrating factors in particular classes of differential equations.

If $M\,dx + N\,dy = 0$ is given equation, then we find the integrating factors by following rules.

$\boxed{\textbf{Theorem 3}}$ Suppose that $M dx + N dy = 0$ is non-exact differential equation.

(a) If $\dfrac{M_x - N_y}{N} = f(x)$, a function of x alone then $\mu = e^{\int f(x)\,dx}$ is an integrating factor for the equation $M dx + N dy = 0$.

(b)   If $\dfrac{N_x - M_y}{M} = g(y)$ is a function y alone then $\mu = e^{\int g(y)\, dy}$ is an integrating factor for $Mdx + Ndy = 0$.

Proof : Suppose $\mu(x, y)$ is an integrating factor for the equation

$$Mdx + Ndy = 0.$$

Then $\mu Mdx + \mu Ndy = 0$ is exact.

$$\Rightarrow \quad \frac{\partial}{\partial y}(\mu M) = \frac{\partial}{\partial x}(\mu N) \quad \Rightarrow \quad \mu\frac{\partial M}{\partial y} + M\frac{\partial \mu}{\partial y} = \mu\frac{\partial N}{\partial x} + N\frac{\partial \mu}{\partial x}.$$

Now, if μ is a function of x alone, then $\dfrac{\partial \mu}{\partial y} = 0$, and $\dfrac{\partial \mu}{\partial x} = \dfrac{d\mu}{dx}$, so that

$$\frac{1}{\mu}\frac{d\mu}{dx} = \frac{1}{N}\left(\frac{\partial M}{\partial y} - \frac{\partial N}{\partial x}\right) \quad \Rightarrow \quad \frac{d\mu}{\mu} = \frac{1}{N}(M_y - N_x)\, dx$$

Integrating this equation, we get

$$\mu = e^{\int f(x)\, dx}, \text{ where } f(x) = \frac{1}{N}(M_y - N_x).$$

Similarly, if μ is a function a function y alone, then we get,

$$\mu = e^{\int g(y)\, dy}, \text{ where } g(y) = \frac{1}{M}(N_x - M_y). \qquad \blacksquare$$

Proposition : Suppose there exists functions f and g such that

$$M_y - N_x = f(x)\, N - g(y)\, M.$$

Then, $\mu(x, y) = P(x)\, Q(y)$, is an integrating factor for $Mdx + Ndy = 0$ where $P(x) = \pm\, e^{\int f(x)\, dx}$ and $Q(y) = \pm\, e^{\int g(y)\, dy}$.

Proof : Suppose that $\mu(x, y) = P(x)\, Q(y)$.

Consider $\dfrac{\partial}{\partial y}(\mu M) = \dfrac{\partial}{\partial x}(\mu N)$.

$$\Leftrightarrow \quad \frac{\partial}{\partial y}(P(x)\, Q(y)\, M) = \frac{\partial}{\partial x}(P(x)\, Q(y)\, N)$$

$$\Leftrightarrow \quad P(x)\, Q'(y)\, M + P(x)\, Q(y)\, M_y = P'(x)\, Q(y)\, N + P(x)\, Q(y)\, N_x$$

$$\Leftrightarrow \quad P(x)\, Q'(y)\, M - P'(x)\, Q(y)\, N = P(x)\, Q(y)\, (N_x - M_y)$$

$$\Leftrightarrow \quad \frac{P'(x)}{P(x)}\, N - \frac{Q'(y)}{Q(y)}\, M = M_y - N_x$$

Let $f(x) = \dfrac{P'(x)}{P(x)}$ and $g(y) = \dfrac{Q'(y)}{Q(y)}$.

Then $\log P(x) = \int f(x)\, dx,\ \log Q(y) = \int g(y)\, dy$

$\Rightarrow\quad P(x) = \pm\, e^{\int f(x)\, dx}$ and $Q(y) = \pm\, e^{\int g(y)\, dy}$

Then, $M_y - N_x = f(x)\, N - g(y)\, M.$

Thus, if f and g are functions such that $M_y - N_x = f(x)\, N - g(y)\, M$, then the function $\mu = P(x)\, Q(y)$ is an integrating factor for

$$Mdx + Ndy = 0, \text{ where, } P(x) = \pm\, e^{\int f(x)\, dx} \text{ and } Q(y) = \pm\, e^{\int g(y)\, dy}$$

Illustrative Examples

Example 4.7 : *Find an integrating factor for*

$$(3xy + 6y^2)\, dx + (2x^2 + 9xy)\, dy = 0 \text{ and solve the equation.}$$

Solution : Here $M = 3x + 6y^2$, $N = 2x^2 + 9xy$ and $M_y - N_x = -x + 3y$

$$\therefore \quad \frac{M_y - N_x}{M} = \frac{-x + 3y}{3xy + 6y^2}, \quad \frac{N_x - N_y}{N} = \frac{x - 3y}{2x^2 + gxy}$$

Now, we find the functions $f(x)$ and $g(y)$ such that

$$M_y - N_x = f(x)\, N - g(y)\, M$$

$$\Rightarrow\quad -x + 3y = f(x)\,(2x^2 + 9xy) - g(y)\,(3xy + 6y^2)$$

$$\Rightarrow\quad -x + 3y = x\, f(x)\,(2x + 9y) - yg(y)\,(3x + 6y)$$

Comparing coefficients on L.H.S. and R.H.S, we have

$$x\, f(x) = A,\ \ yg(y) = B \text{ are constants.}$$

$$\Rightarrow \qquad\qquad -x + 3y = A(2x + gy) - B(3x + 6y)$$

$$\qquad\qquad\qquad = (2A - 3B)\, x + (9A - 6B)\, y$$

$$\Rightarrow \qquad\qquad 2A - 3B = 1 \ \ \text{and}$$

$$\qquad\qquad\qquad 9A - 6B = 3$$

$$\Rightarrow\quad A = 1 \ \text{ and }\ B = 1$$

Therefore, $\qquad x\, f(x) = 1 \Rightarrow f(x) = \dfrac{1}{x} \Rightarrow P(x) = \pm\, e^{\int f(x)\, dx} = \pm\, x$

and $\qquad\qquad yg(y) = 1 \Rightarrow g(y) = \dfrac{1}{y} \Rightarrow Q(y) = \pm\, y$

$\Rightarrow\quad$ Integrating factor is $\mu(x, y) = P(x)\, Q(y) = xy.$

Multiplying by μ, the equation becomes

$$(3x^2y^2 + 6xy^3)\, dx + (2x^3y + gx^2y^2)\, dy = 0 \text{ which is exact.}$$

It is general solution is $x^3y^2 + 3x^2y^3 = c.$

Example 4.8 : *Solve the equation* $- ydx + (x + x^6)\, dy = 0.$

Solution : Here $M = -y$, $N = x + x^6$ and $M_y - N_x = -2 - 6x^5$

Let $f(x)$ and $g(y)$ be the functions such that

$$M_y - N_x = f(x)\, N - g(y)\, M$$

$$\Rightarrow \quad -2 - 6x^5 = f(x)\,(x + x^6) + g(y)\, y$$

$$\Rightarrow \quad f(x) = \frac{-6}{x} \text{ and } g(y) = \frac{4}{y}$$

$$\Rightarrow \quad P(x) = \pm\, e^{\int f(x)\, dx} = \pm\frac{1}{x^6} \text{ and } Q(y) = \pm\, y^4$$

$\therefore$ Integrating factor is $\mu(x, y) = P(x)\, Q(y) = \dfrac{y^4}{x^6}$.

$\therefore$ $\mu(Mdx + Ndy) = 0$ is exact.

$$\Rightarrow \quad \frac{-y^5}{x^6}\, dx + \left(\frac{y^4}{x^5} + y^4\right) dy = 0 \text{ is exact.}$$

$\therefore$ Solution is $\left(\dfrac{y}{x}\right)^5 + y^5 = c.$

Proposition : Suppose a, b, c and d are constants with $ad - bc \neq 0$, and let m and n be arbitrary real numbers. Then $\mu(x, y) = x^\alpha\, y^\beta$ is an integrating factor for the differential equation.

$$(ax^m y + by^{n+1})\, dx + (cx^{m+1} + dxy^n)\, dy = 0.$$

Proof : Consider the equation

$$x^\alpha\, y^\beta\, (ax^m y + by^{n+1})\, dx + x^\alpha\, y^\beta\, (cx^{m+1} + dxy^n)\, dy = 0 \quad \ldots (1)$$

Equation (1) will be exact if

$$\frac{\partial}{\partial y}\, (x^\alpha\, y^\beta\, (ax^m y + by^{n+1})) = \frac{\partial}{\partial x}\, (x^\alpha\, y^\beta\, (cx^{m+1} + dxy^n))$$

$$\Rightarrow \quad (\beta + 1)\, ax^{m+\alpha}\, y^\beta + b(n + \beta + 1)\, x^\alpha\, y^{n+\beta} = c(\alpha + m + 1)\, x^{\alpha + m}\, y^\beta$$

$$+ (\alpha + 1)\, dx^\alpha y^{\beta + n}$$

Comparing coefficients of $x^{m+\alpha}\, y^\beta$ and $x^\alpha\, y^{n+\beta}$ on L.H.S. and R.H.S. we get,

$$a(\beta + 1) = c(\alpha + m + 1) \text{ and}$$

$$b(n + \beta + 1) = d(\alpha + 1)$$

$$\Rightarrow \quad c\alpha - a\beta = a\beta - c(m + 1) \quad \ldots (2)$$

$$\text{and} \quad d\alpha - b\beta = b(n + 1) - d \quad \ldots (3)$$

Since, $ad - bc \neq 0$, there exists unique $\alpha, \beta \in \mathbb{R}$ that satisfies (2) and (3). Simultaneously.

If (α, β) satisfies (2) and (3) simultaneously, then equation (1) is exact. Therefore, $\mu = x^{\alpha} y^{\beta}$ is an integrating factor for the given differential equation.

Example 4.9 : *Consider the differential equation*

$$ydx + \left(2x + \frac{1}{y}\right) dy = 0 \qquad \qquad \text{... (1)}$$

and $\quad y^2 dx + (2xy + 1) dy = 0 \qquad \qquad \text{... (2)}$

(a) *Verify that $\mu(x, y) = y$ is an integrating factor for equations (1) and (2).*

(b) *Verify that $y \equiv 0$ is a solution of (2) but not of (1).*

(c) *Show that $y(xy + 1) = c$ is an implicit solution of (2) and explain why every differentiable function other than $y \equiv 0$ that satisfies $y(xy + 1) = c$ is also a solution of (1).*

Solution : (a) If we multiply equation (1) by $\mu(x, y) = y$, then we get

$$y^2 dx + y \left(2x + \frac{1}{y}\right) dy = 0$$

$$\Rightarrow \qquad y^2 dx + (2xy + 1) dy = 0$$

This is an exact equation.

$\Rightarrow \quad \mu(x, y) = y$ is an integrating factor.

(b) Now, $y \equiv 0$ satisfies (2), but $y \equiv 0$ do not satisfy (1).

(c) The general solution of (2) is

$$xy^2 + y = c \Rightarrow y(xy + 1) = c,$$

hence, it is a solution of (1).

Now, if $y = y_1(x)$ is a function that satisfies $y(xy + 1) = c$, then $y_1(xy_1 + 1) = c$. Differentiating w.r.t. x, we get

$$y_1^2 + 2xy_1 \frac{dy_1}{dx} + \frac{dy_1}{dx} = 0$$

$$\Rightarrow \qquad (2xy_1 + 1) \frac{dy_1}{dx} = - y_1^2$$

$$\Rightarrow \qquad y_1^2 dx + (2xy_1 + 1) dy = 0$$

$\therefore \quad y_1$ satisfies the given differential equation (1), since

$$y_1 \left(y_1 dx + \left(2x + \frac{1}{y_1}\right) dy_1\right) = 0.$$

Rules of Finding Integrating Factors (I.F.) :

Rule - 1 : If $Mx + Ny \neq 0$ and the given equation is homogeneous, then $\dfrac{1}{Mx + Ny}$ is an integrating factor of the equation.

Note : When $Mx + Ny = 0$, then $\dfrac{M}{N} = -\dfrac{y}{x}$, then the differential equation $M\,dx + N\,dy = 0$ i.e. $\dfrac{dy}{dx} = \dfrac{-M}{N}$ reduces to $\dfrac{dy}{dx} = \dfrac{y}{x} \Rightarrow \dfrac{dy}{y} = \dfrac{dx}{x}$ in which variables are separated.

Integrating we get

$$\log y = \log x + c_1$$

$$\Rightarrow \qquad \log\left(\dfrac{y}{x}\right) = \log c$$

$$\Rightarrow \qquad \dfrac{y}{x} = c$$

$$\Rightarrow \qquad y = cx \qquad\qquad \text{where, } k = \pm c$$

which is a general solution of equation $M\,dx + N\,dy = 0$, when $Mx + Ny = 0$.

Rule - 2 : If $Mx - Ny \neq 0$ and the equation $M\,dx + N\,dy = 0$ has the form

$$f_1(xy)\, y\, dx + f_2(xy)\, x\, dy = 0,$$

then $\dfrac{1}{Mx - Ny}$ is an integrating factor of the equation.

Note : If $Mx - Ny = 0$, then $\dfrac{M}{N} = \dfrac{y}{x}$. In this case the equation $M\,dx + N\,dy = 0$ reduces to

$$\dfrac{dy}{dx} = -\dfrac{y}{x}$$

$$\Rightarrow \qquad \dfrac{dx}{x} + \dfrac{dy}{y} = 0$$

Integrating we get $\qquad xy = c$

which is the general solution of $M\,dx + N\,dy = 0$, when

$$Mx - Ny = 0.$$

Rule - 3 : If $\dfrac{\dfrac{\partial M}{\partial y} - \dfrac{\partial N}{\partial x}}{N}$ is a function of x alone, say f(x) or constant c, then

$$\text{I.F.} \;=\; e^{\int f(x)\,dx} \;\; \text{or} \;\; e^{\int c\,dx}$$

of an equation   M dx + N dy = 0

Rule - 4 : If $\dfrac{\dfrac{\partial N}{\partial x} - \dfrac{\partial M}{\partial y}}{M}$ is a function of y alone, say f(y) or a constant c, then

$$\text{I.F.} \;=\; e^{\int f(y)\,dy} \;\; \text{or} \;\; e^{\int c\,dy}$$

of the equation M dx + N dy = 0.

Rule - 5 : If an equation is of the form

$$x^a y^b \,(my\,dx + nx\,dy) + x^c y^d \,(py\,dx + qx\,dy) = 0$$

where a, b, c, d, p, q, m and n are all constants then $x^\alpha\, y^\beta$ is an integrating factor of the equation, where α and β are so determined that when the equation is multiplied by $x^\alpha\, y^\beta$ the resulting equation is exact.

Illustrative Examples

Example 4.10 : *Solve,* $(x^2 y - 2xy^2)\,dx - (x^3 - 3x^2 y)\,dy = 0.$

Solution : The given equation is

$$(x^2 y - 2xy^2)\,dx - (x^3 - 3x^2 y)\,dy = 0 \qquad\qquad \dots \text{(i)}$$

Comparing with M dx + N dy = 0, we have

$$M = x^2 y - 2xy^2 \;\; \text{and} \;\; N = -(x^3 - 3x^2 y)$$

$$\Rightarrow \quad \frac{\partial M}{\partial y} = x^2 - 4xy \;\; \text{and} \;\; \frac{\partial N}{\partial x} = -(3x^2 - 6xy)$$

$$\Rightarrow \quad \frac{\partial M}{\partial y} \neq \frac{\partial N}{\partial x}$$

$\therefore$ The given equation is not exact.

Here we observe that M and N are homogeneous functions of x and y of same degree 3. Hence given equation (i) is homogeneous differential equation.

Hence its integrating factor is $\dfrac{1}{Mx + Ny}$

$$\therefore \qquad \text{I.F.} = \frac{1}{Mx + Ny} = \frac{1}{x^2 y^2} \neq 0$$

Now, multiplying the given equation (i) by I.F. $\dfrac{1}{x^2 y^2}$ we get

$$\left(\frac{1}{y} - \frac{2}{x}\right) dx + \left(\frac{3}{y} - \frac{x}{y^2}\right) dy = 0$$

which is an exact differential equation. Hence its general solution is

$$\int \left(\frac{1}{y} - \frac{2}{x}\right) dx + \int \frac{3}{y} \, dy = c$$

y-constant

$$\Rightarrow \qquad \frac{x}{y} - 2 \log x + 3 \log y = c$$

$$\Rightarrow \qquad \frac{x}{y} + \log \left(\frac{y^3}{x^2}\right) = c$$

Example 4.11: *Solve,* $(x^2 y^2 + xy + 1)\, y\, dx + (x^2 y^2 - xy + 1)\, x\, dy = 0$

Solution : The given equation is

$$(x^2 y^2 + xy + 1)\, y\, dx + (x^2 y^2 - xy + 1)\, x\, dy = 0 \qquad\qquad \dots \text{(i)}$$

which is not exact.

The given equation is of the form.

$$f_1(xy)y \, dx + f_2(xy)x \, dy = 0$$

Hence, its I.F. is $\dfrac{1}{Mx - Ny}$

Here, $\qquad\qquad M = (x^2 y^2 + xy + 1)y \ \text{ and } \ N = (x^2 y^2 - xy + 1)x$

$$\therefore \qquad Mx - Ny = (x^2 y^2 + xy + 1)xy - (x^2 y^2 - xy + 1)\, xy$$

$$= 2x^2 y^2 \neq 0$$

$$\therefore \qquad\qquad \text{I.F.} = \frac{1}{2x^2 y^2}$$

Multiplying the given equation (i) by I.F. $\dfrac{1}{2x^2 y^2}$, it becomes

$$\frac{1}{2}\left(y + \frac{1}{x} + \frac{1}{x^2 y}\right) dx + \frac{1}{2}\left(x - \frac{1}{y} + \frac{1}{xy^2}\right) dy = 0$$

which can be shown to be an exact and whose general solution is obtained as usual and we get it as

$$xy - \frac{1}{xy} + \log \left(\frac{x}{y}\right) = c$$

Example 4.12 : *Solve, $(x^2 + y^2 + 2x)\, dx + 2y\, dy = 0.$*

Solution : The given differential equation is

$$(x^2 + y^2 + 2x)\, dx + 2y\, dy = 0 \qquad \text{... (i)}$$

Comparing with M dx + N dy = 0, we have

$$M = x^2 + y^2 + 2x \text{ and } N = 2y$$

$$\Rightarrow \quad \frac{\partial M}{\partial y} = 2y \text{ and } \frac{\partial N}{\partial x} = 0$$

$$\Rightarrow \quad \frac{\partial M}{\partial y} \neq \frac{\partial N}{\partial x}$$

Hence, the given equation is not exact equation.

But we observe that

$$\frac{\dfrac{\partial M}{\partial y} - \dfrac{\partial N}{\partial x}}{N} = \frac{2y - 0}{2y} = 1 = f(x) \text{ say}$$

$$\therefore \qquad \text{I.F.} = e^{\int f(x)\, dx} = e^{\int 1\, dx} = e^{x}$$

Multiplying equation (i) by I.F., we get

$$e^{x}(x^2 + y^2 + 2x)\, dx + 2ye^{x}\, dy = 0$$

which is exact and its general solution is

$$\int e^{x}(x^2 + y^2)\, 2x\, dx + \int 0\, dy = c$$
$$\text{y-constant}$$

$$\Rightarrow \qquad e^{x}(x^2 + y^2) = c$$

Example 4.13 : *Solve, $(xy^3 + y)\, dx + 2(x^2y^2 + x + y^4)\, dy = 0.$*

Solution : The given differential equation is

$$(xy^3 + y)\, dx + 2(x^2y^2 + x + y^4)\, dy = 0 \qquad \text{... (i)}$$

Comparing with M dx + N dy = 0, we get

$$M = xy^3 + y \text{ and } N = 2(x^2y^2 + x + y^4)$$

$$\Rightarrow \quad \frac{\partial M}{\partial y} = 3xy^2 + 1 \text{ and } \frac{\partial N}{\partial x} = 2(2xy^2 + 1)$$

$$\Rightarrow \quad \frac{\partial M}{\partial y} \neq \frac{\partial N}{\partial x}$$

$$\therefore \quad \text{Equation (i) is not exact. But we observe that}$$

$$\frac{\dfrac{\partial N}{\partial x} - \dfrac{\partial M}{\partial y}}{M} = \frac{2(2xy^2 + 1) - (3xy^2 + 1)}{xy^3 + y}$$

$$= \frac{xy^2 + 1}{y(xy^2 + 1)}$$

$$= \frac{1}{y} = f(y) \text{ say}$$

$$\therefore \qquad \text{I.F.} = e^{\int f(y)\, dy}$$

$$= e^{\int 1/y\, dy} = e^{\log y} = y$$

$$\therefore \qquad \text{I.F.} = y$$

Multiplying the given equation (i) by I.F., we get

$$(xy^4 + y^2)\, dx + 2(x^2y^3 + xy + y^5)dy = 0$$

which is an exact equation. Hence its solution is

$$\int_{y\text{-constant}} (xy^4 + y^2)\, dx + \int 2y^5\, dy = c$$

$$\Rightarrow \qquad \frac{1}{2} x^2 y^4 + xy^2 + \frac{1}{3} y^6 = c$$

Example 4.14 : *Solve,* $(y^2 + 2x^2y)\, dx + (2x^3 - xy)\, dy = 0.$

Solution : The given differential equation is

$$(y^2 + 2x^2y)\, dx + (2x^3 - xy)\, dy = 0 \qquad \qquad \dots \text{(i)}$$

which can be rewritten as

$$y(y + 2x^2)\, dx + 2x^2\, (y\, dx + x\, dy)$$

Let us suppose that the possible I.F. be $x^\alpha y^\beta$.

$\therefore$ Multiplying equation (i) by I.F. $x^\alpha y^\beta$, we have

$$(x^\alpha y^{\beta+2} + 2x^{\alpha+2} y^{\beta+1})\, dx + (2x^{\alpha+3} y^\beta - x^{\alpha+1} y^{\beta+1})\, dy = 0 \ \dots \text{(ii)}$$

Comparing equation (ii) with $M\, dx + N\, dy = 0$, we get

$$M = x^\alpha y^{\beta+2} + 2x^{\alpha+2} y^{\beta+1} \text{ and } N = 2x^{\alpha+3} y^\beta - x^{\alpha+1} y^{\beta+1}$$

$$\Rightarrow \quad \frac{\partial M}{\partial y} = (\beta + 2)\, x^\alpha y^{\beta+1} + 2(\beta + 1)\, x^{\alpha+2} y^\beta$$

$$\text{and } \frac{\partial N}{\partial x} = 2(\alpha + 3)\, x^{\alpha+2} y^\beta - (\alpha + 1)\, x^\alpha y^{\beta+1}$$

If the equation (ii) is exact, then

$$\frac{\partial M}{\partial y} = \frac{\partial N}{\partial x}$$

$$\Rightarrow \quad (\beta + 2)\, x^{\alpha}\, y^{\beta + 1} + 2(\beta + 1)\, x^{\alpha + 2}\, y^{\beta} = 2(\alpha + 3)\, x^{\alpha + 2}\, y^{\beta}$$
$$- (\alpha + 1)\, x^{\alpha}\, y^{\beta + 1}$$

Equating coefficients of $x^{\alpha}\, y^{\beta + 1}$ and $x^{\alpha + 2}\, y^{\beta}$, we have

$$\beta + 2 = -(\alpha +) \Rightarrow \alpha + \beta = -3$$
$$2(\beta + 1) = 2(\alpha + 3) \Rightarrow \alpha - \beta = -2$$

Solving these equations, we get

$$\alpha = -\frac{5}{2} \quad \text{and} \quad \beta = -\frac{1}{2}$$

For these values of α and β equation (ii) becomes

$$\left(x^{-\frac{5}{2}}\, y^{\frac{3}{2}} + 2x^{-\frac{1}{2}}\, y^{\frac{1}{2}} \right) dx + \left(2x^{\frac{1}{2}}\, y^{-\frac{1}{2}} - x^{-\frac{3}{2}}\, y^{\frac{1}{2}} \right) dy = 0$$

which is an exact equation. Hence its general solution is

$$\int_{y\text{-constant}} \left(x^{-\frac{5}{2}}\, y^{\frac{3}{2}} + 2x^{-\frac{1}{2}}\, y^{\frac{1}{2}} \right) dx + \int 0\, dy = c$$

$$\Rightarrow \qquad -\frac{2}{3} x^{-\frac{3}{2}}\, y^{\frac{3}{2}} + 4x^{\frac{1}{2}}\, y^{\frac{1}{2}} = c$$

which is required general solution.

Example 4.15 : *Solve,*

[cos x tan y + cos (x + y)] dx + [sin x sec²y + cos (x + y)] dy = 0

Solution : Comparing the given equation with M dx + N dy = 0, we get

$$M = \cos x \tan y + \cos (x + y) \text{ and } N = \sin x \sec^2 y + \cos (x + y)$$

$$\Rightarrow \quad \frac{\partial M}{\partial y} = \cos x \sec^2 y - \sin (x + y) \text{ and } \frac{\partial N}{\partial x} = \cos x \sec^2 y - \sin (x + y)$$

$$\Rightarrow \quad \frac{\partial M}{\partial y} = \frac{\partial N}{\partial x}$$

$\Rightarrow$ the given differential equation is exact differential equation.

Hence its general solution is

$$\int_{y\text{-constant}} M\, dx + \int (\text{terms in N free from x})\, dy = c$$

$$\Rightarrow \qquad \int_{y\text{-constant}} [\cos x \tan y + \cos (x + y)]dx = c$$

$$\Rightarrow \quad \sin x \tan y + \sin (x + y) = c$$

which is the required general solution.

Example 4.16 : *Solve,* $(x^2 - ay)\, dx - (ax - y^2)\, dy = 0$

Solution : Comparing the given equation with M dx + N dy = 0, we get

$$M = x^2 - ay \quad \text{and} \quad N = -(ax - y^2)$$

$$\Rightarrow \quad \frac{\partial M}{\partial y} = -a \quad \text{and} \quad \frac{\partial N}{\partial x} = -a$$

$$\Rightarrow \quad \frac{\partial M}{\partial y} = \frac{\partial N}{\partial x}$$

$\Rightarrow$　the given differential equation is exact differential equation

Hence the required general solution of the given equation is

$$\int M\, dx + \int (\text{terms in N free from x})\, dy = c_1$$

y-constant

$$\Rightarrow \quad \int (x^2 - ay)dx + \int y^2 dy = c_1$$

y-constant

$$\Rightarrow \quad \frac{1}{3} x^3 - axy + \frac{1}{3} y^3 = c_1$$

$$\Rightarrow \quad x^3 - 3\, axy + y^3 = c_1, \qquad\qquad \text{where, } c = 3c_1$$

Example 4.17 : *Solve,* $(x^4 + y^4)\, dx - xy^3\, dy = 0$

Solution : The given equation is

$$(x^4 + y^4)\, dx - xy^3\, dy = 0 \qquad\qquad \dots \text{(i)}$$

Comparing equation (i) with M dx + N dy = 0, we get

$$M = x^4 + y^4 \quad \text{and} \quad N = -xy^3$$

$$\Rightarrow \quad \frac{\partial M}{\partial y} = 4y^3 \quad \text{and} \quad \frac{\partial N}{\partial x} = -y^3$$

$$\Rightarrow \quad \frac{\partial M}{\partial y} \neq \frac{\partial N}{\partial x}$$

$\therefore$　The given equation is not exact.

Here we observe that given equation is homogeneous equation.

Hence its I.F. is $\dfrac{1}{Mx + Ny}$ we have,

$$\begin{aligned}
Mx + Ny &= (x^4 + y^4)\, x + (-xy^3)\, y \\
&= x^5 + xy^4 - xy^4 \\
&= x^5 \neq 0
\end{aligned}$$

$$\therefore \qquad\qquad Mx + Ny = x^5$$

$\therefore$ $$\text{I.F.} = \frac{1}{Mx + Ny} = \frac{1}{x^5}$$

Now multiplying the given equation (i) by I.F. $\frac{1}{x^5}$, we get

$$\left(\frac{1}{x} + \frac{y^4}{x^5}\right) dx - \frac{y^3}{x^4} dy = 0$$

which is an exact differential equation. Hence its general solution is

$$\int_{\text{y-constant}} \left(\frac{1}{x} + \frac{y^4}{x^5}\right) dx + \int 0\, dy = c$$

$\Rightarrow$ $$\log x - \frac{y^4}{4x^4} = c$$

which is required general solution.

Example 4.18 : *Solve, $y(xy + 2x^2y^2)\, dx + x(xy - x^2y^2)\, dy = 0$*

Solution : The given equation is

$$y(xy + 2x^2y^2)\, dx + x(xy - x^2y^2)\, dy = 0 \qquad \ldots \text{(i)}$$

which is not exact.

The given equation is of the form

$$f_1(xy)\, y\, dx + f_2(xy)\, x\, dy = 0$$

Hence its I.F. is $\dfrac{1}{Mx - Ny}$

Here, $$M = y(xy + 2x^2y^2) \text{ and } N = x(xy - x^2y^2)$$

$\therefore$ $$\begin{aligned} Mx - Ny &= xy(xy + 2x^2y^2) - xy\,(xy - x^2y^2) \\ &= x^2y^2 + 2x^3y^3 - x^2y^2 + x^3y^3 \\ &= 3x^3y^3 \neq 0 \end{aligned}$$

$\therefore$ $$Mx - Ny = 3x^3y^3$$

$\therefore$ $$\text{I.F.} = \frac{1}{Mx - Ny} = \frac{1}{3x^3y^3}$$

Now multiplying the given equation (i) by I.F. $\frac{1}{3x^3y^3}$, we get

$$\frac{y(xy + 2x^2y^2)}{3x^3y^3}\, dx + \frac{x(xy - x^2y^2)}{3x^3y^3}\, dy = 0$$

$\Rightarrow$ $$\left(\frac{1}{3x^2y} + \frac{2}{3x}\right) dx + \frac{1}{3}\left(\frac{1}{xy^2} - \frac{1}{y}\right) dy = 0$$

which is exact. Hence its general solution is

$$\int_{y\text{-constant}} \left(\frac{1}{3x^2y} + \frac{2}{3x}\right) dx + \int -\frac{1}{3y} \, dy = c$$

$$\Rightarrow \quad \log\left(\frac{x^2}{y}\right) - \frac{1}{xy} = c$$

which is required general solution.

Example 4.19 : *Solve,*

$$[xy \sin(xy) + \cos(xy)]y \, dx + [xy \sin(xy) - \cos(xy)]x \, dy = 0$$

(April 2013)

Solution : The given differential equation is

$$[xy \sin(xy) + \cos(xy)]y \, dx + [xy \sin(xy) - \cos(xy)]x \, dy = 0 \quad \dots \text{(i)}$$

which is not exact.

The given equation is of the form

$$f_1(xy)y \, dx + f_2(xy)x \, dy = 0$$

Hence its I.F. is $\dfrac{1}{Mx - Ny}$

Here, $\quad M = [xy \sin(xy) + \cos(xy)]y$ and

$$N = [xy \sin(xy) - \cos(xy)]x$$

$$\therefore \quad Mx - Ny = xy\,[xy \sin(xy) + \cos(xy)] -$$

$$xy[xy \sin(xy) - \cos(xy)]$$

$$= x^2y^2 \sin(xy) + xy \cos(xy) - x^2y^2 \sin(xy) + xy \cos(xy)$$

$$= 2xy \cos(xy) \neq 0$$

$$\therefore \quad \text{I.F.} = \frac{1}{Mx - Ny} = \frac{1}{2xy \cos(xy)}$$

Multiplying the given equation (i) by I.F. $\dfrac{1}{2xy \cos(xy)}$, we get

$$\frac{1}{2}\left[\tan(xy) + \frac{1}{xy}\right] y \, dx + \frac{1}{2}\left[\tan(xy) - \frac{1}{xy}\right] x \, dy = 0$$

which can be shown to be exact. Hence its general solution is

$$\int_{y\text{-constant}} \frac{1}{2}\left[\tan(xy) + \frac{1}{xy}\right] y \, dx + \int 0 \, dy = c_1$$

$$\Rightarrow \qquad \int \left[y \tan (xy) + \frac{1}{x} \right] dx = c_2 \qquad (\because \ 2c_1 = c_2)$$

$$\Rightarrow \qquad \log \sec (xy) + \log x = \log c_2$$

$$\therefore \qquad x \sec (xy) = c_2$$

Example 4.20 : *Solve,* $(x^3 + xy^4)\,dx + 2y^3\,dy = 0$

Solution : The given differential equation is

$$(x^3 + xy^4)\,dx + 2y^3\,dy = 0 \qquad \qquad \dots \text{(i)}$$

Comparing this equation with M dx + N dy = 0, we get

$$M = x^3 + xy^4 \ \text{ and } \ N = 2y^3$$

$$\Rightarrow \quad \frac{\partial M}{\partial y} = 4xy^3 \quad \text{and} \quad \frac{\partial N}{\partial x} = 0$$

$$\Rightarrow \quad \frac{\partial M}{\partial y} \neq \frac{\partial N}{\partial x}$$

$\therefore$ Given differential equation is not exact equation.

But we observe that

$$\frac{\dfrac{\partial M}{\partial y} - \dfrac{\partial N}{\partial x}}{N} = \frac{4xy^3 - 0}{2y^3} = 2x = f(x)$$

$\therefore$ I.F. $= e^{\int f(x)\,dx} = e^{\int 2x\,dx} = e^{x^2}$.

$\therefore$ Multiplying given equation by I.F. e^{x^2}, we get

$$e^{x^2}(x^3 + xy^4)\,dx + 2e^{x^2}\,dy = 0$$

which is exact and its general solution is

$$\int_{\text{y-constant}} e^{x^2}(x^3 + xy^4)\,dx + \int 0\,dy = c_1$$

$$\Rightarrow \qquad \int e^{x^2} x^3\,dx = y^4 \int e^{x^2} x\,dx = c_1$$

Put $x^2 = t \Rightarrow 2x\,dx = dt$

$$\Rightarrow \qquad \frac{1}{2}\int te^t\,dt + \frac{1}{2}y^4 \int e^t\,dt = c_1$$

Integrating by parts, we get

$$\Rightarrow \qquad \frac{1}{2}\left[te^t - \int e^t\,dt\right] + \frac{1}{2}y^4 e^t = c_1$$

$$\Rightarrow \qquad \frac{1}{2}\left[te^t - e^t\right] + \frac{1}{2}y^4 e^t = c_1$$

$$\Rightarrow \qquad (t - 1)\, e^t + y^4\, e^t \;=\; 2c_1$$

$$\Rightarrow \qquad (x^2 - 1)\, e^{x^2} + y^4\, (e^{x^2}) \;=\; 2c_1$$

$$\Rightarrow \qquad (x^2 + y^4 - 1)\, e^{x^2} \;=\; c \qquad \text{where, } 2c_1 = c$$

which is required general solution.

Example 4.21 : *If $f(x)$ a function of x only is an integrating factor of the differential equation.*

$$(x^4 e^x - 2mxy^2)\, dx + 2mx^2 y\, dy \;=\; 0$$

Find the form of $f(x)$ and hence solve the equation.

Solution : The given differential equation is

$$(x^4 e^x - 2mxy^2)\, dx + 2m\, x^2 y\, dy \;=\; 0 \qquad \text{... (i)}$$

We have given $f(x)$ is an I.F. of given equation (i).

$\therefore$ Multiplying given equation by $f(x)$ we should have the exact differential equation.

$$(x^4 e^x - 2mxy^2)\, f(x)\, dx + 2mx^2 y\, f(x)\, dy = 0 \qquad \text{... (ii)}$$

Comparing equation (ii) with $M\, dx + N\, dy = 0$, we have

$$M = (x^4 e^x - 2mxy^2)\, f(x) \quad \text{and} \quad N = 2mx^2 y\, f(x)$$

$$\Rightarrow \quad \frac{\partial M}{\partial y} = -\,4xy\, f(x) \quad \text{and} \quad \frac{\partial N}{\partial x} = 2my\, [x^2 f'(x) + 2x\, f(x)]$$

Since equation (ii) is exact.

$$\therefore \qquad \frac{\partial M}{\partial y} \;=\; \frac{\partial N}{\partial x}$$

$$\Rightarrow \qquad -\,4mxy\, f(x) \;=\; 2my\, [x^2 f'(x) + 2x\, f(x)]$$

$$\Rightarrow \qquad xf'(x) + 4f(x) \;=\; 0$$

$$\Rightarrow \qquad \frac{f'(x)}{f(x)} \;=\; -\,\frac{4}{x}$$

Integrating, we get

$$\log f(x) \;=\; -\,4 \log x$$

$$\Rightarrow \qquad \log f(x) \;=\; \log\!\left(\frac{1}{x^4}\right)$$

$$\Rightarrow \qquad f(x) \;=\; \frac{1}{x^4}$$

$\therefore$ The integrating factor $f(x) = \dfrac{1}{x^4}$ using this I.F. the equation (ii) becomes

$$(x^4 e^x - 2mxy^2)\dfrac{1}{x^4}\,dx + 2mx^2 y\left(\dfrac{1}{x^4}\right)dy = 0$$

$$\Rightarrow \qquad \left(e^x - \dfrac{2my^2}{x^3}\right)dx + \left(\dfrac{2my}{x^2}\right)dy = 0$$

which is an exact equation and hence its general solution is

$$\int\left[e^x - 2m\dfrac{y^2}{x^3}\right]dx + \int 0\,dy = c$$

y-constant

$$\Rightarrow \qquad \int e^x\,dx - 2my^2\int x^{-3}\,dx = c$$

$$\Rightarrow \qquad e^x - 2my^2\left(-\dfrac{1}{2x^2}\right) = c$$

$$\Rightarrow \qquad e^x + m\dfrac{y^2}{x^2} = c$$

$$\Rightarrow \qquad x^2 e^x + my^2 = cx^2$$

which is required solution.

Example 4.22 : *Given that the integrating factor of the equation*

$$y\,sec^2x\,dx + \left[3\tan x - \left(\dfrac{\sec y}{y}\right)^2\right]dy = 0$$

is of the form y^n. Find n and solve the equation.

Solution : The given differential equation is

$$y\,sec^2x\,dx + \left[3\tan x - \left(\dfrac{\sec y}{y}\right)^2\right]dy = 0 \qquad\qquad \dots\text{(i)}$$

We have given that the integrating factor of equation (i) is of the form y^n.

$\therefore$ Multiplying the equation (i) by I.F. y^n, we should have exact differential equation.

$$y^{n+1}\sec^2x\,dx + [3y^n\tan x - y^{n-2}\sec^2y]\,dy = 0 \qquad\qquad \dots\text{(ii)}$$

Here, $M = y^{n+1}\sec^2x$ and $N = 3y^n\tan x - y^{n-2}\sec^2y$

$$\Rightarrow \quad \dfrac{\partial M}{\partial y} = (n+1)\,y^n\sec^2x \quad\text{and}\quad \dfrac{\partial N}{\partial x} = 3y^n\sec^2x$$

Since equation (ii) is exact.

$$\therefore \qquad \frac{\partial M}{\partial y} = \frac{\partial N}{\partial x}$$

$$\Rightarrow \qquad (n + 1)\, y^n \sec^2 x = 3y^n \sec^2 x$$

$$\Rightarrow \qquad n + 1 = 3 \Rightarrow n = 2$$

$\therefore$ y^2 is an I.F. of the given equation. Using $n = 2$ in equation (ii), we get

$$y^3 \sec^2 x\, dx + [3y^2 \tan x - \sec^2 y]\, dy = 0$$

which is exact. Hence, its general solution is

$$\int_{y\text{-constant}} y^3 \sec^2 x\, dx + \int (-\sec^2 y)\, dy = c$$

$$\Rightarrow \qquad y^3 \tan x - \tan y = c$$

which is required general solution.

Think Over It

1. Consider the system of first order ordinary differential equations :

$$\begin{cases} \dfrac{dx}{dt} = ax + by \\[2mm] \dfrac{dy}{dt} = cx + dy \end{cases}$$

   (a) How to solve this system ?

   (b) What are phase portraits of this system ?

2. Given second order ordinary differential equation

$$y'' + P(x)y' + Q(x)y = R(x),$$

how to formulate it as a system of first order ordinary differential equations ?

3. How to use differential equations in following ?

   (a) Biological Modelling.

   (b) Mechanical Vibration Analysis.

   (c) Fluid Dynamics and Heat Transfer.

   (d) Earthquake Analysis.

   (e) Control Theory.

Summary

1. Differential equation $Mdx + Ndy = 0$ is exact if there exists a function $u(x, y)$ with continuous second order partial derivatives such that

$$du = \frac{\partial u}{\partial x}\,dx + \frac{\partial u}{\partial y}\,dy = Mdx + Ndy$$

2. Differential equation $Mdx + Ndy = 0$ is exact if and only if $\dfrac{\partial M}{\partial y} = \dfrac{\partial N}{\partial x}$

3. A function $\mu(x, y)$ is an integrating factor for $Mdx + Ndy = 0$, if $\mu Mdx + \mu Ndy = 0$ is exact.

4. If $\dfrac{1}{N}\left(\dfrac{\partial M}{\partial y} - \dfrac{\partial N}{\partial x}\right) = f(x)$, a function of x alone, then $\mu = e^{\int f(x)\,dx}$ is an integrating factor for $Mdx + Ndy = 0$.

5. If $\dfrac{1}{M}\left(\dfrac{\partial N}{\partial x} - \dfrac{\partial M}{\partial y}\right) = g(y)$, a function of y alone, then $\mu = e^{\int g(y)\,dy}$ is an integrating factor for $Mdx + Ndy = 0$.

6. If the equation $Mdx + Ndy = 0$ is homogeneous and $Mx + Ny \neq 0$, then $\dfrac{1}{Mx + Ny}$ is an integrating factor for $Mdx + Ndy = 0$.

7. If the equation $Mdx + Ndy = 0$ has the form

   $f_1(xy)\,ydx + f_2(xy)\,xdy$ and $Mx - Ny \neq 0$, then its integrating factor is $\dfrac{1}{M_x - N_y}$.

8. If the equation is in the form

$$x^a y^b\,(Mydx + Nxdy) + x^c y^d\,(pydx + qxdy) = 0,$$

   then its integrating factor is $\mu(x, y) = x^\alpha y^\beta$.

9. If there exists functions f and g such that

$$\frac{\partial M}{\partial y} - \frac{\partial N}{\partial x} = f(x)\,N - g(y)\,M,$$ then there are functions $P(x)$ and $Q(y)$ such that $\mu = P(x)\,Q(y)$ is an integrating factor for $Mdx + Ndy = 0$.

Exercise

[A] Say True or False : Justify !

1. Every differential equation which has a general solution is exact.

2. If the differential equation Mdx + Ndy = 0 has an integrating factor, then it has many integrating factors.

3. Every differential has an integrating factor.

4. If the differential equation Mdx + Ndy = 0 has a general solution, then it has an integrating factor.

5. if the differential equation is variable separable, then it has no integrating factor.

[B] Multiple Choice Questions : Choose the Correct Alternative

1. Integrating factor for the equation $\dfrac{dy}{dx} = f(x)\,g(y)$ is ……

   (a) x

   (b) $\dfrac{1}{x}$

   (c) $\dfrac{1}{f(x)\,g(y)}$

   (d) $\dfrac{1}{g(y)}$

2. If the differential equation Mdx + Ndy = 0 is exact, then its integrating factor is ……

   (a) 0

   (b) 1

   (c) x

   (d) y

3. If the differential equation Mdx + Ndy = 0 is exact and $f(x, y) = c$ is its general solution, then ……

   (a) $\dfrac{dy}{dx} = \dfrac{-f_x}{f_y}$

   (b) $\dfrac{dy}{dx} = f(x, y)$

   (c) $\dfrac{dy}{dx} = \dfrac{\partial f}{\partial y}$

   (d) $\dfrac{dy}{dx} = 0$

4. Which of the following differential equation is exact ?

   (a) $dx + dy = 0$

   (b) $2y\,dx + 3x\,dy = 0$

   (c) $x\,dx + dy = 0$

   (d) $dx + y\,dy = 0$

5. Which of the following differential equation is exact ?

   (a) $(1 - F_x)\,dx - F_y\,dy = 0$

   (b) $F_y\,dx + F_x\,dy = 0$

   (c) $F_x\,dx - F_y\,dy = 0$

   (d) $- F_x\,dx + F_y\,dy = 0$

6. Differential equation for the family $f(x, y) = c$ is ......

   (a) $\dfrac{dy}{dx} = f(x, y)$ (b) $\dfrac{dy}{dx} = \dfrac{f_x}{f_y}$

   (c) $\dfrac{dy}{dx} = \dfrac{-f_x}{f_y}$ (d) $\dfrac{dy}{dx} = \dfrac{-f_y}{f_x}$

7. Integrating factor for the differential equation $ydx + xdy = 0$ is ...

   (a) $\dfrac{x}{y}$ (b) y^2

   (c) $\dfrac{1}{y^2}$ (d) $\dfrac{x^2}{y^2}$

8. The number of integrating factors for a differential equation of the type $M\,dx + N\,dy = 0$ is ......

   (a) one   (b) finite

   (c) infinite   (d) none of these

9. If the given equation of the type

   $$f_1(xy)y\,dx + f_2(xy)x\,dy = 0$$

   then its I.F. is ......

   (a) $Mx - Ny$ (b) $\dfrac{1}{Mx - Ny}$

   (c) $Mx + Ny$ (d) $\dfrac{1}{Mx + Ny}$

10. If the equation $M\,dx + N\,dy$ is homogeneous and $Mx + Ny \neq 0$, then its I.F. is ......

    (a) $Mx - Ny$ (b) $Mx + Ny$

    (c) $\dfrac{1}{Mx + Ny}$ (d) $\dfrac{1}{Mx - Ny}$

11. In the differential equation $Mdx + Ndy = 0$, if $\dfrac{1}{M}\left(\dfrac{\partial N}{\partial x} - \dfrac{\partial M}{\partial y}\right)$ is a function of y alone say $f(y)$, then required integrating factor is ......

    (a) $f(y)$ (b) $e^{\int f(y)\,dy}$

    (c) $\int f(y)\,dy$ (d) none of these

12. In the differential equation $Mdx + Ndy = 0$, if $\dfrac{1}{N}\left(\dfrac{\partial N}{\partial y} - \dfrac{\partial M}{\partial x}\right)$ is a function of x alone say $f(x)$, then required integrating factor is ......

    (a) $f(x)$ (b) $\int f(x)\,dx$

    (c) $e^{\int f(x)\,dx}$ (d) none of these

[C] Theory Questions :

1. Define exact differential equation.

2. State and prove the necessary and sufficient condition for exactness of the differential equation $Mdx + Ndy = 0$.

3. If $u(x, y)$ has continuous first order partial derivatives then $u(x, y) = c$ is the general solution of the differential equation $u_x dx + u_y dy = 0$.

4. Suppose that M, N are functions such that their first order partial derivatives are continuous on an open rectangle R. Let G be a function such that $\dfrac{dG}{dx} = M$. If $M_y \neq N_x$ in R, then prove that

$$\frac{\partial}{\partial x}\left(N - \frac{\partial G}{\partial y}\right) \neq 0.$$

5. If the equations $M_1 dx + N_1 dy = 0$ and $M_2 dx + N_2 dy = 0$ are exact, then prove that the equation

$(M_1 + M_2)\, dx + (N_1 + N_2)\, dy = 0$ is also exact.

6. Suppose that all second order partial derivatives of M and N are continuous and $Mdx + Ndy = 0$, $-Ndx + Mdy = 0$ are exact on an open rectangle R.

Then prove that $M_{xx} + M_{yy} = 0$ and $N_{xx} + N_{yy} = 0$ on R.

7. Suppose that all second order partial derivatives of $u(x, y)$ are continuous and u is harmonic function on an open rectangle R, then prove that the equation $-u_y dx + u_x dy = 0$ is exact on R.

Moreover, prove that there is a function v such that $v_x = -u_y$ and $v_y = u_x$ on R.

8. Prove that if a differential equation $Mdx + Ndy = 0$ has a general solution, then it has an integrating factor.

9. If $\dfrac{1}{N}\left(\dfrac{\partial M}{\partial y} - \dfrac{\partial N}{\partial x}\right) = f(x)$ is a function of a single variable x, then prove that $e^{\int f(x)\, dx}$ is an integrating factor for $Mdx + Ndy = 0$.

10. If $\dfrac{1}{M}\left(\dfrac{\partial N}{\partial x} - \dfrac{\partial M}{\partial y}\right) = g(y)$, a function of single variable y, then prove that $e^{\int g(y)\, dy}$ is an integrating factor for $Mdx + Ndy = 0$.

11. Suppose that there exists functions $f(x)$ and $g(y)$ such that $M_y - N_x = f(x) N - g(y) M$. Then prove that $\mu = e^{\int f(x)\, dx} \cdot e^{\int g(y)\, dy}$ is an integrating factor for $Mdx + Ndy = 0$.

12. Suppose that a, b, c, d are constants with $ad - bc \neq 0$ and let m, n be arbitrary real numbers. Then prove that $\mu = x^\alpha\, y^\beta$ is an integrating factor for the differential equation

$$(ax^m y + by^{n+1})\, dx + (cx^{m+1} + dxy^n)\, dy = 0.$$

[D] Numerical Problems :

1. Solve the equation

$$(ye^{xy} \tan x + e^{xy} \sec^2 x)\, dx + xe^{xy} \tan x\, dy = 0$$

2. Determine whether the following differential equations are exact, hence solve them

   (i) $6x^2 y^2\, dx + 4x^3 y\, dy = 0$

   (ii) $(3y \cos x + 4xe^x + 2x^2 e^x)\, dx + (3 \sin x + 3)\, dy = 0$

   (iii) $14x^2 y^3\, dx + 21x^2 y^2\, dy = 0$

   (iv) $(2x - 2y^2)\, dx + (12y^2 - 4xy)\, dy = 0$

   (v) $(x + y)^2\, dx + (x + y)^2\, dy = 0$

   (vi) $(- 2y^2 \sin x + 3y^3 - 2x)\, dx + (4y \cos x + 9xy^2)\, dy = 0$

   (vii) $(3x^2 + 2xy + 4y^2)\, dx + (x^2 + 8xy + 18y)\, dy = 0$

   (viii) $\left(\dfrac{1}{x} + 2x\right) dx + \left(\dfrac{1}{y} + 2y\right) dy = 0$

   (ix) $(x^2 e^{x^2 + y} (2x^2 + 3) + 4x)\, dx + (x^3 e^{x^2 + y} - 12y^2)\, dy = 0$

   (x) $(3x^2 \cos (xy) - x^3 y \sin (xy) + 4x)\, dx + (8y - x^4 (\sin xy))\, dy = 0$

3. Solve the following initial value problems :

   (i) $(4x^3 y^2 - 6x^2 y - 2x - 3)\, dx + (2x^4 y - 2x^3)\, dy = 0,\ y(1) = 3$

   (ii) $(- 4y \cos x + 4 \sin x \cos x + \sec^2 x)\, dx + (4y - \sin x)\, dy = 0,\ y\left(\dfrac{\pi}{4}\right) = 0$

   (iii) $(y^3 - 1) e^x\, dx + 3y^2 (e^x + 1)\, dy = 0,\ y(0) = 0$

   (iv) $(\sin x - y \sin x - 2 \cos x)\, dx + \cos x\, dy = 0,\ y(0) = 1$

4.   (a)   Solve the equation $(x^2 + y^2)\, dx + 2xy\, dy = 0$ implicitly.

     (b)   For what choices of (x_0, y_0) does the initial value problem $(x^2 + y^2)\, dx + 2xy\, dy = 0$, $y(x_0) = y_0$ has a unique solution $y = y(x)$ on some open interval (a, b) that contains x_0 ?

     (c)   Plot a direction field and some integral curves for this equation.

5.   Find all functions M such that the equation is exact :

     (i) $M(x, y)\, dx + (x^2 - y^2)\, dy = 0$

     (ii) $M(x, y)\, dx + 2xy \sin x \cos y\, dy = 0$

     (iii) $M(x, y)\, dx + (e^x - e^y \sin x)\, dy = 0$

6.   Find all functions N such that the equation is exact :

     (i) $(x^3 y^2 + 2xy + 3y^2)\, dx + N(x, y)\, dy = 0$

     (ii) $(ln(xy) + 2y \sin x)\, dx + N(x, y)\, dy = 0$

     (iii) $(x \sin x + y \sin y)\, dx + N(x, y)\, dy = 0$

7.   Suppose that M and N are continuous and have continuous partial derivatives M_y and N_x that satisfy the condition $M_y = N_x$ on an open rectangle R. Show that if $(x, y) \in R$ and

$$F(x, y) = \int_{x_0}^{x} M(s, y_0)\, ds + \int_{y_0}^{y} N(x, t)\, dt, \text{ then } F_x = M \text{ and } F_y = N.$$

8.   Solve the initial value problem

$$y' + \frac{2}{x}y = -\frac{2xy}{x^2 + 2x^2 y + 1}, \; y(1) = -2.$$

9.   Solve the initial value problem

$$y' + 2xy = -e^{-x^2}\left(\frac{3x + 2ye^{x^2}}{2x + 3ye^{x^2}}\right), \; y(0) = -1.$$

10.   Verify that the following functions are harmonic and find their harmonic conjugates :

     (i) $e^x \cos y$,   (ii) $x^3 - 3xy^2$,

     (iii) $\cos x \cos h\, y$,   (iv) $\sin x \cos h\, y$.

11. (a) Verify that $\mu(x, y) = \dfrac{1}{(x - y)^2}$ is an integrating factor for $- y^2 dx + x^2 dy = 0$ on an open rectangle that does not intersect the line $y = x$.

   (b) Show that $\dfrac{xy}{x - y} = c$ is an implicit solution of the given equation.

12. Find integrating factor and solve the following equations :

   (i) $ydx - xdy$

   (ii) $3x^2y\, dx + 2x^3\, dy = 0$

   (iii) $2y^3\, dx + 3y^2\, dy = 0$

   (iv) $(5xy + 2y + 5)\, dx + 2x\, dy = 0$

   (v) $(xy + x + 2y + 1)\, dx + (x + 1)\, dy = 0$

   (vi) $(27xy^2 + 8y^3)\, dx + (18x^2y + 12xy^2)\, dy = 0$

   (vii) $(6xy^2 + 2y)\, dx + (12x^2y + 6x + 3)\, dy = 0$

   (viii) $y^2dx + \left(xy^2 + 3xy + \dfrac{1}{y}\right) dy = 0$

   (ix) $(12x^3y + 24x^2y^2)\, dx + (9x^4 + 32x^3y + 4y)\, dy = 0$

   (x) $- ydx + (x^4 - x)\, dy = 0$

   (xi) $\cos x \cos y\, dx + (\sin x \cos y - \sin x \sin y + y)\, dy = 0$

13. Find an integrating factor of the form $\mu(x, y) = P(x)\, Q(y)$ for the following differential equations and solve them :

   (i) $y(1 + 5\, ln|x|)\, dx + 4x\, ln|x|\, dy = 0$

   (ii) $(\alpha y + \gamma xy)\, dx + (\beta x + \delta xy)\, dy = 0$

   (iii) $(3x^2y^3 - y^2 + y)\, dx + (- xy + 2x)\, dy = 0$

   (iv) $2y\, dx + 3(x^2 + x^2y^3)\, dy = 0$

   (v) $(a \cos (xy) - y \sin (xy))\, dx + (b \cos xy) - x \sin (xy)\, dy = 0$

   (vi) $x^4y^4\, dx + x^5y^3\, dy = 0$

14. Solve the following differential equations :

   (i) $(3x^2y^4 + 2xy)\, dx + (2x^3y^3 - x^2)\, dy = 0$

   (ii) $(x^2y + y^3)\, dx + \left(\dfrac{2}{3}x^3 + 4xy^2\right) dy = 0$

(iii) $(2xy^4e^y + 2xy^3 + y)\, dx + (x^2y^4e^y - x^2y^2 - 3x)\, dy = 0$

(iv) $(x - y^2)\, dx + 2xy\, dy = 0$

(v) $\left(y + \dfrac{y^3}{3} + \dfrac{x^2}{2}\right) dx + \dfrac{1}{4}(x + xy^2)\, dy = 0$

(vi) $(x^2 + y^2)\, dx - 4xy\, dy = 0$

(vii) $(3x + 2y^2)\, y\, dx + 2(2x + 3y^2)\, x\, dy = 0$

(viii) $(2x^2y - 3y^4)\, dx + (3x^3 + 2xy^3)\, dy = 0$

(ix) $(y^2 + 2x^2y)\, dx + (2x^3 - xy)\, dy = 0$

(x) $(2y\, dx + 3x\, dy) + 2xy\,(3y\, dx + 4x\, dy) = 0$

(xi)   Find r, if x^r is an I.F. of the equation.

$(x + y^3)\, dx + 6xy^2\, dy = 0$

Answers

[A] (1) False   (2) True   (3) False   (4) True   (5) False

[B] (1) - (d)   (2) - (b)   (3) - (a)   (4) - (a)   (5) - (a)   (6) - (c)   (7) - (c)
(8) - (c)   (9) - (b)   (10) - (c)   (11) - (b)   (12) - (c)

[D]

14. (i)   **Hint :** I.F. $= \dfrac{1}{y^2}$

Ans. : $x^3y^3 + x^2 = cy$

(ii)   **Hint :** I.F. $= y$

Ans. : $\dfrac{1}{3}x^3y^2 + xy^4 = c$

(iii) **Hint :** I.F. $= \dfrac{1}{y^4}$

Ans. : $x^2e^y + \dfrac{x^2}{y} + \dfrac{x}{y^3} = c$

(iv)   **Hint :** I.F. $= \dfrac{1}{x^2}$

Ans. : $\log x + \dfrac{y^2}{x} = c$

(v)   **Hint :** I.F. $= x^3$

     Ans. : $3x^4y + x^4y^3 + x^6 = c$

(vi) **Hint :** I.F. $= x^{-\frac{3}{2}}$

     Ans. : $\dfrac{2}{3} x^{\frac{3}{2}} - 2\dfrac{y^2}{\sqrt{x}} = c$

(vii) **Hint :** $\alpha = 1, \ \beta = 1$

     Ans. : $x^3y^4 + x^2y^6 = c$

(viii) **Hint :** $\alpha = -\dfrac{49}{13}, \ \beta = -\dfrac{28}{13}$

     Ans. : $5x^{-\frac{36}{13}} y^{\frac{24}{13}} - 12 \ x^{-\frac{10}{13}} y^{-\frac{15}{13}} = c$

(ix) **Hint :** $\alpha = -\dfrac{5}{2}, \ \beta = -\dfrac{1}{2}$

     Ans. : $4x^{\frac{1}{2}} y^{\frac{1}{2}} + \dfrac{2}{3} x^{-\frac{3}{2}} y^{\frac{3}{2}} = c$

(x)   **Hint :** $\alpha = 1, \ \beta = 2$

     Ans. : $x^2y^3 + 2x^3y^4 = c$

(xi) **Hint :** $r = -\dfrac{1}{2}$

     Ans. : $\dfrac{2}{3} x^{\frac{3}{2}} + 2x^{\frac{1}{2}} y^3 = c$

☞ ☞ ☞

Appendix

Practical Problems

(A) Derivative

Q.1. Show that the function

$$f(x) = x^2 \;\; ; \;\; \text{if} \;\; x \geq 0$$
$$= -x^2 \;\; ; \;\; \text{if} \;\; x < 0$$

is differentiable on $\mathbb{R}$

Q.2. Let

$$f(x) = \frac{x\,(e^{-1/x} - e^{1/x})}{e^{-1/x} + e^{1/x}}, \; x \neq 0$$
$$= 0 \qquad\qquad , \; x = 0$$

Discuss the continuity and differentiability of f at the origin.

Q.3. Let $f(x) = |x - 2| + |x| + |x + 2|, \; x \in \mathbb{R}$

Discuss the differentiability of f on $\mathbb{R}$.

Q.4. If

$$f(x) = x^2 \sin\left(\frac{1}{x}\right), \; x \neq 0$$
$$= 0 \qquad , \; x = 0$$

Then show that f is differentiable at every $x \in \mathbb{R}$, but the derivative is not continuous at $x = 0$.

Q.5. Let $f(x) = \begin{cases} 1 + x & , \; x \leq 0 \\ x & , \; 0 < x < 1 \\ 2 - x & , \; 1 \leq x \leq 2 \\ 3x - x^2 & , \; x > 2 \end{cases}$

Discuss the continuity and differentiability of f on $\mathbb{R}$.

Q.6. Let $f(x) = \begin{cases} \dfrac{1}{2}(b^2 - a^2) & , \; \text{for } 0 \leq x \leq a \\ \dfrac{1}{2}b^2 - \dfrac{x^2}{6} - \dfrac{a^3}{3x} & , \; \text{for } a < x \leq b \\ \dfrac{1}{3}\left(\dfrac{b^3 - a^3}{x}\right) & , \; \text{for } x > 3 \end{cases}$

Prove that f and f' are continuous but f'' discontinuous.

(B) Mean Value Theorem

Q.1. If $y = \sin^{-1} x$, prove that $y' = \dfrac{1}{\sqrt{1 - x^2}}$, $\forall\, x \in \mathbb{R}$

Q.2. If $y = \sec^{-1} x$, prove that $y' = \dfrac{1}{|x|\,\sqrt{x^2 - 1}}$.

Q.3. Let $g(x) = f(x) + f(1 - x)$ and $f''(x) > 0$, $\forall\, x \in \mathbb{R}$.

Find the interval of increasing and decreasing of $g(x)$.

Q.4. Show that :

(i)　$x(1 + x) < \log(1 + x) < x$, $\forall\, x > 0$

(ii)　$x - \dfrac{x^2}{2} < \log(1 + x) < x - \dfrac{x^2}{2(1 + x)}$, $\forall\, x > 0$

(iii)　$\dfrac{x}{1 + x} < \log(1 + x) < x$, $x > 0$

(iv)　$\dfrac{\tan x}{x} > \dfrac{x}{\sin x}$ for $0 < x < \dfrac{\pi}{2}$

(v)　$\dfrac{2}{\pi} < \dfrac{\sin x}{x} < 1$, $0 < x < \dfrac{\pi}{2}$

Q.5. If $0 < a, b < 1$, prove that $\dfrac{b - a}{\sqrt{1 - a^2}} < \sin^{-1} b - \sin^{-1} a < \dfrac{b - a}{\sqrt{1 - b^2}}$.

Q.6. Verify Cauchy's Mean Value theorem for $f(x) = e^x$, $g(x) = e^{-x}$ on $[a, b]$. Also, find the values of c.

Q.7. Let $a_0, a_1, a_2, \ldots, a_n$ be real numbers, such that

$$\frac{a_n}{n + 1} + \frac{a_{n-1}}{n} + \ldots + \frac{a_1}{2} + a_0 = 0$$

Prove that there is a root of the polynomial

$$a_n x^n + a_{n-1} x^{n-1} + \ldots + a_1 x + a_0 \text{ in } [0, 1]$$

Q.8. Verify Cauchy's Mean Value theorem for $f(x) = \dfrac{1}{x^2}$, $g(x) = \dfrac{1}{x}$ on $[a, b]$, where $0 < a < b$. Also, find the value of c.

(C) Successive Differentiation

Q.1. If $p^2 = a^2 \cos^2 \theta + b^2 \sin^2 \theta$, prove that $p + \dfrac{d^2 p}{d\theta^2} = \dfrac{a^2 b^2}{p^3}$.

Q.2. If $y = \left[\log\left(\dfrac{x + \sqrt{x^2 - ab}}{a}\right)\right]^2 + < \log(x + \sqrt{x^2 - a^2})$, then prove that

$$(x^2 - a^2)\dfrac{d^2 y}{dx^2} + x\dfrac{dy}{dx} = 2a.$$

Q.3. Find the nth derivative of $y = \dfrac{x}{x^2 + a^2}$.

Q.4. If $y = x \log\left(\dfrac{x-1}{x+1}\right)$, prove that

$$\dfrac{d^n y}{dx^n} = (-1)^n \, (n-2)! \left[\dfrac{x-n}{(x+1)^n} - \dfrac{x+n}{(x+1)^n}\right]$$

Q.5. If $y = e^{m \sin^{-1} x}$.

Q.6. $y = \cos(m \sin^{-1} x)$. Show that

$$(1 - x^2)\, y_{n+2} - (2n+1)\, xy_{n+1} + (m^2 - n^2)\, y_n = 0$$

Q.7. If $y = (x^2 - 1)^n$, prove that $(x^2 - 1)\, y_{n+2} + 2xy_{n+1} - n(n+1)\, y_n = 0$

Q.8. Let u_n denotes the nth derivative of $\dfrac{(L_x + M)}{(x^2 - 2B_x + c)}$. Prove that

$$\dfrac{x^2 - 2B_x + c}{(n+1)(n+2)}\, u_{n+2} + \dfrac{2(x-B)}{n+1}\, u_{n+1} + u_n = 0$$

Q.9. Obtain the expression of $\log \cosh x$ in powers of x by Maclaurin's theorem.

Q.10. If $y = \sqrt{1 - x^2}\, \sin^{-1} x$, prove that :

   (a) $y_3(1 - x^2) - 3y_2\, x + 2 = 0$

   (b) $y_{n+3}(1 - x^2) - (2n+3)\, x\, y_{n+2} - n(n+2)\, y_{n+1} = 0$

   (c) $\sqrt{(1 - x^2)}\, \sin^{-1} x = x - \dfrac{x^3}{3} - \dfrac{2}{3}\dfrac{x^5}{5} - \dfrac{2 \cdot 4}{3 \cdot 5}\dfrac{x^7}{7} + \cdots$

   (d) $\theta \cot \theta = 1 - \dfrac{\sin^2 \theta}{3} - \dfrac{2}{3}\dfrac{\sin^4 \theta}{5} - \dfrac{2 \cdot 4}{3 \cdot 5}\dfrac{\sin^6 \theta}{7} - \cdots$

   (e) $\dfrac{\pi}{4} = 1 - \dfrac{1}{3}\left(\dfrac{1}{2}\right) - \dfrac{2}{3}\left(\dfrac{1}{5}\right)\left(\dfrac{1}{2}\right)^2 - \dfrac{2}{3}\cdot\dfrac{4}{3}\cdot\dfrac{1}{7}\cdot\left(\dfrac{1}{2}\right)^3 + \cdots$

Q.11. Expand $2x^3 + 7x^2 + x - 6$ in powers of $(x - 2)$.

Q.12. Expand $f(x) = \sin x$ in ascending powers of $\left(x - \dfrac{\pi}{2}\right)$.

Q.13. Determine the following limits :

(a) $\displaystyle\lim_{x \to 1} \dfrac{1 + \log x - x}{1 - 2x + x^2}$

(b) $\displaystyle\lim_{x \to 1/2} \dfrac{\cos^2 \pi x}{e^{2x} - 2ex}$

(c) $\displaystyle\lim_{x \to 0} \dfrac{xe^x - \log(x + 1)}{\cos hx - \cos x}$

(d) $\displaystyle\lim_{x \to 0} \dfrac{e^x - e^{\sin x}}{x - \sin x}$

(e) $\displaystyle\lim_{x \to 0} \dfrac{\tan x - x}{x^2 \sin x}$

(f) $\displaystyle\lim_{x \to 0} \dfrac{e^x - e^{-x} - 2 \sin x}{x^3}$

Q.14. If $\displaystyle\lim_{x \to 0} \dfrac{\sin 2x + a \sin x}{x^3}$ is finite, find the value of a.

Q.15. Evaluate following limits :

(a) $\displaystyle\lim_{x \to 0} \left(\dfrac{1}{x} - \dfrac{1}{e^x - 1}\right)$

(b) $\displaystyle\lim_{x \to 0} \left(\dfrac{\tan x}{x}\right)^{1/x}$

(c) $\displaystyle\lim_{x \to 0} \left(2 - \dfrac{x}{a}\right)^{\tan \frac{\pi x}{2a}}$

(d) $\displaystyle\lim_{x \to 0} \left(\dfrac{x - 1}{2x^2} + \dfrac{e^{-x}}{2x \sin hx}\right)$

(D) Ordinary Differential Equation

Q.1. (a) Plot a direction field and some integral curves for

$xy - 2y = -1$ on the rectangular region

$(-1 \le x \le 1, \ -0.5 < y < 1.5)$

(b) Show that the general solution on $(-\infty, 0) \cup (0, \infty)$ is

$$y = \dfrac{1}{2} + cx^2$$

Q.2. Solve the following initial value problems :

(a) $xy' + 2y = 8x^2, \ y(1) = 3$

(b) $(x^2 - 5) y' - 2xy = x(x^2 - 1), \ y(0) = 4$

Q.3. Find the general solution. Also, plot a direction field and some integral curves on the rectangular region $\{-2 \le x \le 2,\ -2 \le y \le 2\}$.

(a) $y' + 3y = 1$

(b) $y' + 2xy = xe^{-x^2}$

(c) $y' + \dfrac{2x}{1 + x^2}\, y = \dfrac{e^{-x}}{1 + x^2}$

(d) $y' + \dfrac{1}{x}\, y = \dfrac{7}{x^2} + 3$

Q.4. Solve the initial value problem and graph the solutions :

(a) $(3y^2 + 4y)\, y' + 4x + \cos x = 0,\ y(0) = 1$

(b) $y' + 2x(y + 1) = 0,\ y(0) = 2$

Q.5. Verify that the function

$$y = \begin{cases} (x^2 - 1)^{5/3} & ,\ -1 < x < 1 \\ 0 & ,\ |x| \ge 1 \end{cases}$$

is a solution of the initial value problem

$$y' = \frac{10}{3}\, xy^{2/5},\ y(0) = -1 \text{ on } (-\infty, \infty).$$

Q.6. Consider the initial value problem

$$y' = 3x(y - 1)^{1/3},\ \ y(x_\infty) = y_0$$

(a) For what points (x_0, y_0) the solution passes through them exists and is unique ?

(b) Find the general solution.

Q.7. (a) Solve the exact equation

$$(x^3 y^4 + 2x)\, dx + (x^4 y^3 + 3y)\, dy = 0$$

(b) What are the points (x_0, y_0) through which

$$(x^3 y^4 + 2x)\, dx + (x^4 y + 3y)\, dy = 0,\ y(x_0) = y_0,$$

has a unique solution on an open interval containing x_0 ?

Q.8. Solve the following initial value problem

$$y' + \frac{2}{x}\, y = \frac{-2xy}{x^2 + 2x^2 y + 1},\ y(1) = -2$$

Q.9. Solve the differential equations :

    (a) $2y^3\,dx + 3y^2\,dy = 0$

    (b) $(5xy + 2y + 5)\,dx + 2x\,dy = 0$

    (c) $(x^4y^3 + y)dx + (x^5y^2 - x)\,dy = 0$

    (d) $(3x^2y^2 + 2y)\,dx + 2x\,dy = 0.$

☞ ☞ ☞

MODEL QUESTION PAPER - I
F.Y.B.Sc. Mathematics : Paper - II
MT - 122, AMG - 2 : Calculus - II

Time : 2 Hours   **Total Marks : 35**

Note :   1.   All questions are compulsory.
           2.   Figures to the right indicate full marks.

Q.1. Attempt any five of the following :   **[5]**

   (a) Find the differential equation for the family of curves $y = cx$.

   (b) Let $f(x) = x^2$, $x \in \mathbb{R}$. Use definition to find the derivative of f at $c \in \mathbb{R}$.

   (c) State the Rolle's mean value theorem for real valued function of a real variable.

   (d) Define linear differential equation of first order.

   (e) Find the conditions on the constants P, Q, R, S, T and U such that the equation

$$(Px^2 + Qxy + Ry^2)\, dx + (Sx^2 + Txy + Uy^2)\, dy = 0 \text{ is exact.}$$

   (f) Evaluate $\lim_{x \to 0} x^x$.

   (g) If $y = (\sin^{-1} x)^2$, prove that $(1 - x^2)\, y'' - xy' - 2 = 0$.

Q.2. Attempt any two of the following :   **[10]**

   (a) If $f : I \to \mathbb{R}$ is strictly increasing continuous function, then prove that the function $f^{-1} : f(I) \to \mathbb{R}$ is differentiable function and $\dfrac{d}{dy} f^{-1}(y) = \dfrac{1}{f'(x)}$, where $f(x) = y$. Hence, find $\dfrac{d}{dy} \sin^{-1} y$.

   (b) State the Taylor's theorem with Lagrange's form of remainder. Hence, expand the function $f(x) = \log(1 + x)$ in ascending powers of x, upto first four terms.

   (c) Define homogeneous function. Further, solve the homogeneous differential equation $(x - y)\, dx + (x + y)\, dy = 0$.

Q.3. Attempt any two of the following :   **[10]**

   (a) State the necessary and sufficient condition for exactness of the differential equation $M(x, y)\, dx + N(x, y)\, dy = 0$.

P.1

Hence, determine whether the differential equation

$$(4x^3 y^3 + 3x^2)\, dx + (3x^4 y^2 + 6y^2)\, dy = 0$$

is exact. Also, solve this differential equation.

(b) Define integrating factor of a differential equation. Find an integrating factor of the differential equation

$$- y\, dx + (x + x^6)\, dy = 0, \text{ and solve it.}$$

(c) Verify Cauchy's mean value theorem for the functions $f(x) = \dfrac{1}{x^2}$

and $g(x) = \dfrac{1}{x}$ in $[a, b]$, the harmonic mean of a and b.

Q.4. Attempt any one of the following : [10]

(a) State and prove the Leibnitz theorem for n^{th} derivative of the product of two differentiable functions. Hence, prove that if $x = \tan(\log y)$, then

$$(1 + x^2)\, y_{n+1} + (2nx - 1)\, y_n + n(n - 1)\, y_{n-1} = 0. \ \forall\, n \in \mathbb{N}.$$

(b) Define Bernoulli's differential equation. Explain the method of solving the Bernoulli's differential equation.

Hence, solve the equation, $3\dfrac{dy}{dx} + \dfrac{2y}{x + 1} = \dfrac{x^3}{y^2}$.

✍ ✍ ✍

MODEL QUESTION PAPER - II
F.Y.B.Sc. Mathematics : Paper - II
MT - 122, AMG - 2 : Calculus - II

Time : 2 Hours   **Total Marks : 35**

Note :   1.   All questions are compulsory.
         2.   Figures to the right indicate full marks.

Q.1. Attempt any five of the following :   **[5]**

(a)   Solve the initial value problem, $y' = y$, $y(0) = 1$.

(b)   Let $f(x) = x^3$, $x \in \mathbb{R}$ and $c \in \mathbb{R}$. Use the definition to find $f'(c)$.

(c)   Find a function f for which $f'(x) = 0$ for all $x \in A \subseteq \mathbb{R}$, but f is not a constant function on A.

(d)   If $y = \sin (ax + b)$, find y_n.

(e)   Determine order and degree of the different equation

$$y = px + \sqrt{a^2 p^2 + b^2}, \ \text{where,} \ p = \frac{dy}{dx}.$$

(f)   Check whether the differential equation

$$(x^2 - 2xy - y^2)\, dx - (x + y)^2\, dy = 0 \text{ is exact.}$$

(g)   Find an integrating factor for the differential equation

$$y\, dx - x\, dy = 0.$$

Q.2. Attempt any two of the following :   **[10]**

(a)   State and prove the Lagrange's mean value theorem.

(b)   Let $y = e^{ax} \sin (bx + c)$, where a, b, c are constants, prove that
$$\frac{d^n y}{dx^n} = (a^2 + b^2)^{n/2}\, e^{ax} \sin (bx + c + n \tan^{-1} (b/a)), \text{ for every } n \in \mathbb{N}.$$

(c)   Reduce the differential equation

$$(2x + y + 3)\, dx \ = \ (2y + x + 1)\, dy$$

to homogeneous differential equation. Hence, solve it by the method of homogeneous equation.

Q.3. Attempt any two of the following :   **[10]**

(a)   Explain the method of solving the differential equation $Mdx + Ndy = 0$, when the variables in it are separable.

(b) Suppose that all second order partial derivatives of u(x, y) are continuous and u is harmonic function on an open rectangle R. Then prove that the differential equation $- u_y dx + u_x dy = 0$ is exact on R.

(c) Prove that $\dfrac{x}{1 + x^2} < \tan^{-1} x < x \; ; \; \forall\, x > 0$

Q.4. Attempt any one of the following :　　　　　　　　　　[10]

(a) State the Maclaurin's theorem with Lagrange's form of remainder. Hence, prove that,

   (i)　$\tan^{-1} x = x - \dfrac{x^3}{3} + \dfrac{x^5}{5} \ldots\ldots$

   (ii)　$\dfrac{\pi}{4} = 1 - \dfrac{1}{3} + \dfrac{1}{5} - \dfrac{1}{7} + \ldots\ldots$

(b) (i)　Explain the method of solving linear first order ordinary differential equation.

   (ii)　Solve the differential equation

$$2(1 + x)\dfrac{dy}{dx} - (1 + 2x)\, y = x^2 \sqrt{1 + x}.$$

☛ ☛ ☛